# MCAT® Critical Analysis and Reasoning Skills (CARS)

2025–2026 Edition: An Annotated Practice Book

Copyright © 2024
On behalf of UWorld, LLC
Dallas, TX
USA

All rights reserved.
Printed in English, in the United States of America.

Reproduction or translation of any part of this work beyond that permitted by Sections 107 and 108 of the United States Copyright Act without the permission of the copyright owner is unlawful.

The Medical College Admission Test (MCAT®) and the United States Medical Licensing Examination (USMLE®) are registered trademarks of the Association of American Medical Colleges (AAMC®). The AAMC® neither sponsors nor endorses this UWorld product.

Facebook® and Instagram® are registered trademarks of Facebook, Inc. which neither sponsors nor endorses this UWorld product.

X is an unregistered mark used by X Corp, which neither sponsors nor endorses this UWorld product.

**Acknowledgments for the 2025–2026 Edition**

Ensuring that the course materials in this book are accurate and up to date would not have been possible without the multifaceted contributions from our team of content experts, editors, illustrators, software developers, and other amazing support staff. UWorld's passion for education continues to be the driving force behind all our products, along with our focus on quality and dedication to student success.

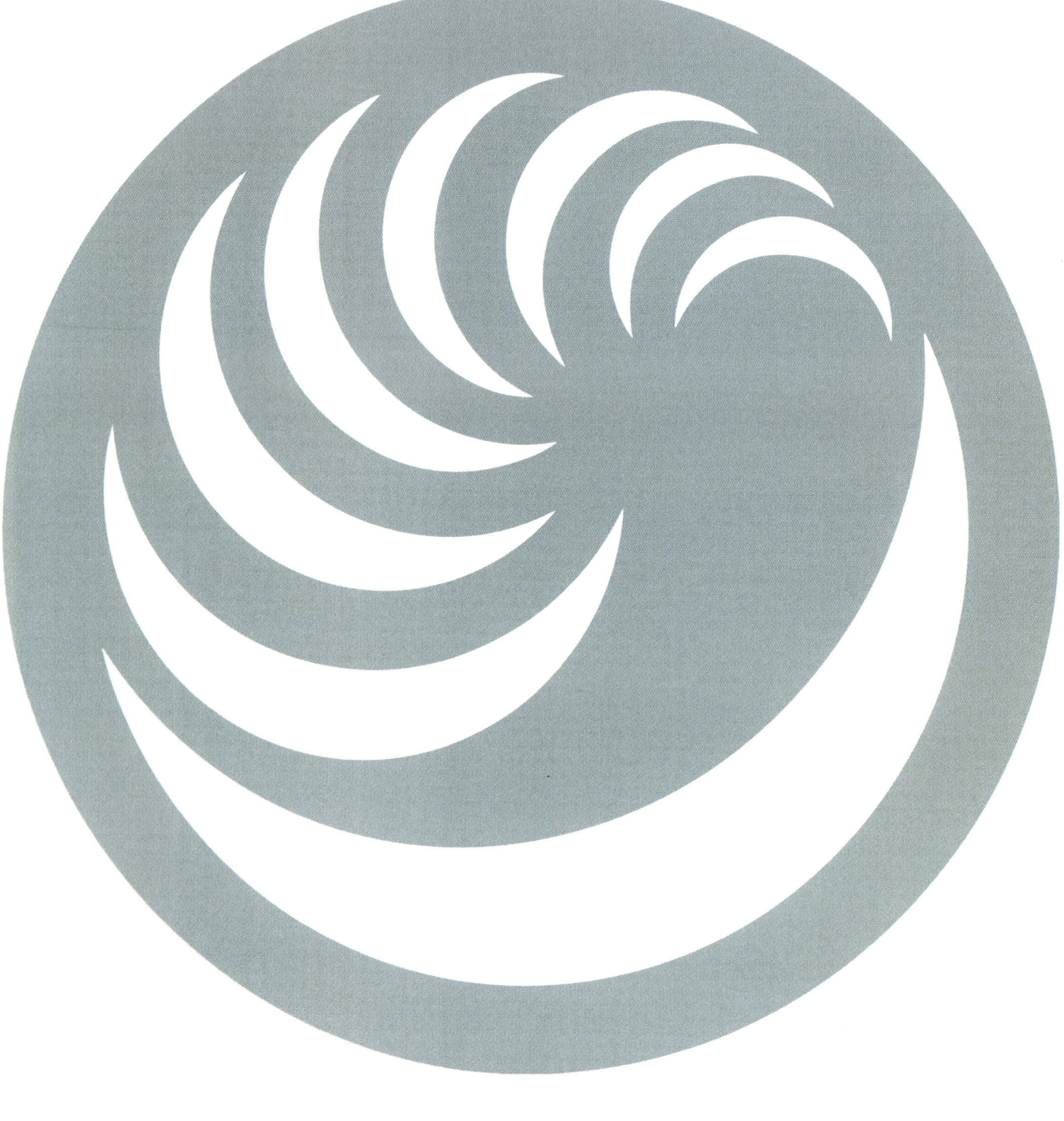

## About the MCAT Exam

Taking the MCAT is a significant milestone on your path to a rewarding career in medicine. Scan the QR codes below to learn crucial information about this exam as you take your next step before medical school.

Basic MCAT Exam Information

Scores and Percentiles

MCAT Sections

Registration Guide

## Preparing for the MCAT with UWorld

The MCAT is a grueling exam spanning seven subjects that is designed to test your aptitude in areas essential for success in medicine. Preparing for the exam can be intimidating—so much so that in post-MCAT questionnaires conducted by the AAMC®, a majority of students report not feeling confident about their MCAT performance.

In response, UWorld set out to create premier learning tools to teach students the entire MCAT syllabus, both efficiently and effectively. Taking what we learned from helping over 90% of medical students prepare for their medical board exams (USMLE®), we launched the UWorld MCAT Qbank in 2017 and the UWorld MCAT UBooks in 2024. The MCAT UBooks are meticulously written and designed to provide you with the knowledge and strategies you need to meet your MCAT goals with confidence and to secure your future in medical school.

Below, we explain how to use the MCAT UBooks and MCAT Qbank together for a streamlined learning experience. By strategically integrating both resources into your study plan, you will improve your understanding of key MCAT content as well as build critical reasoning skills, giving you the best chance at achieving your target score.

## MCAT UBooks: Illustrated and Annotated Guides

The MCAT UBooks include not only the printed editions for each MCAT subject but also provide digital access to interactive versions of the same books. There are eight printed MCAT UBooks in all, six comprehensive review books covering the science subjects and two specialized books for the Critical Analysis and Reasoning Skills (CARS) section of the exam:

- Biology
- Biochemistry
- General Chemistry
- Organic Chemistry
- Physics
- Behavioral Sciences
- CARS (Annotated Practice Book)
- CARS Passage Booklet (Annotated)

Each UBook is organized into Units, which are divided into Chapters. The Chapters are then split into Lessons, which are further subdivided into Concepts.

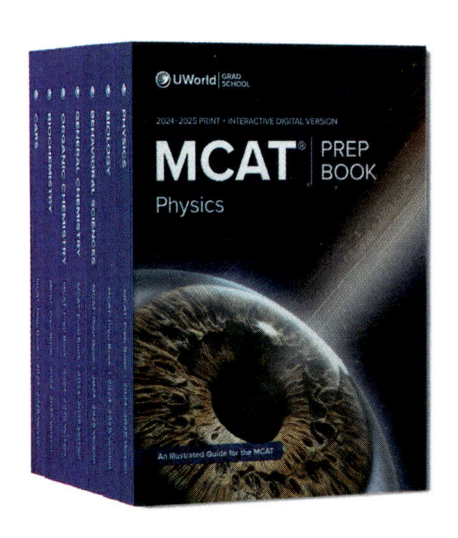

# MCAT Sciences: Printed UBook Features

The MCAT UBooks bring difficult science concepts to life with thousands of engaging, high-impact visual aids that make topics easier to understand and retain. In addition, the printed UBooks present key terms in blue, indicating clickable illustration hyperlinks in the digital version that will help you learn more about a scientific concept.

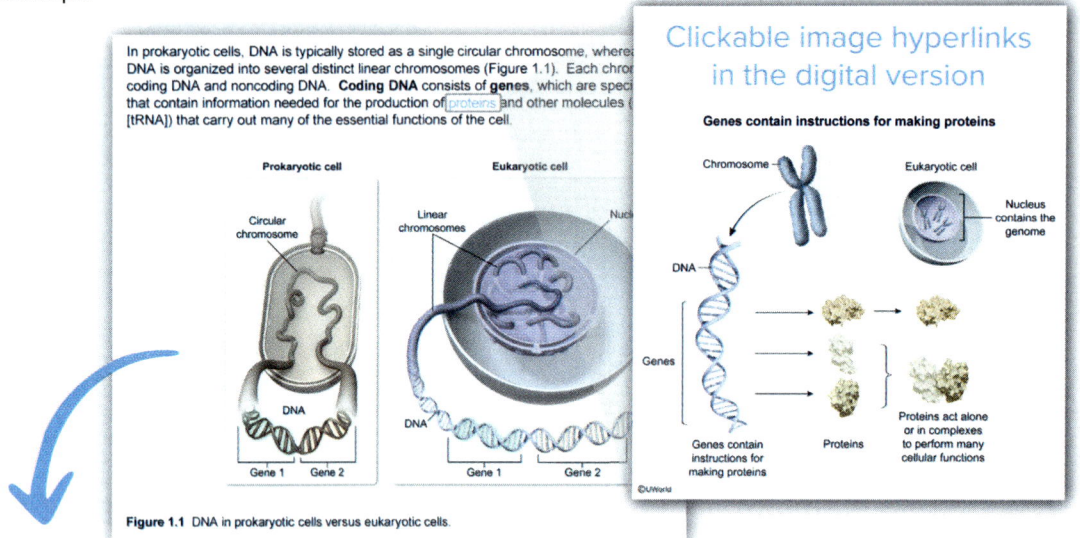

*Thousands of educational illustrations in the print book*

*Clickable image hyperlinks in the digital version*

# Test Your Basic Science Knowledge with Concept Check Questions

The printed UBooks also include 450 new questions—never before available in the UWorld Qbank—for Biology, General Chemistry, Organic Chemistry, Biochemistry, and Physics. These new questions, called Concept Checks, are interspersed throughout the entire book to enhance your learning experience. Concept Checks allow you to instantly test yourself on MCAT concepts you just learned from the UBook.

Short answers to the Concept Checks are found in the appendix at the end of each printed UBook. In addition, the digital version of the UBook provides an interactive learning experience by giving more detailed, illustrated, step-by-step explanations of each Concept Check. These enhanced explanations will help reinforce your learning and clarify any areas of uncertainty you may have.

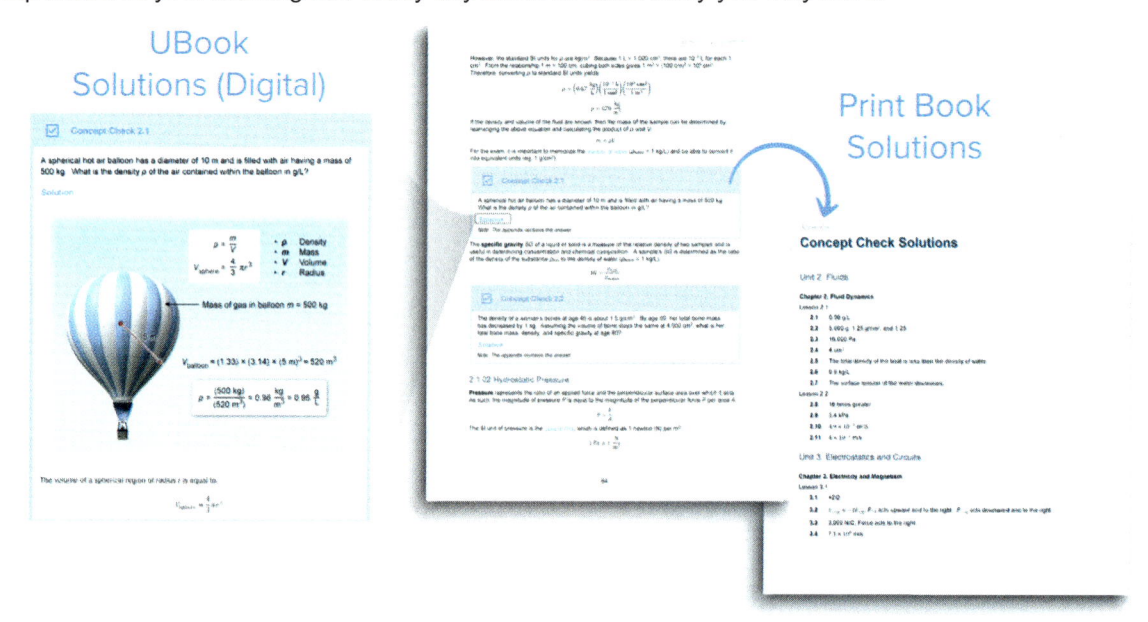

*UBook Solutions (Digital)*

*Print Book Solutions*

# MCAT CARS Printed UBook Features

For CARS, the main book, or Annotated Practice Book, teaches you the specialized CARS skills and strategies you need to master and then follows up with multiple sets of MCAT-level practice questions.

Additionally, the CARS Passage Booklet includes annotated versions of the passages in the CARS Main Book. From these annotations, you will learn how to break down a CARS passage in a step-by-step manner to find the right answer to each CARS question.

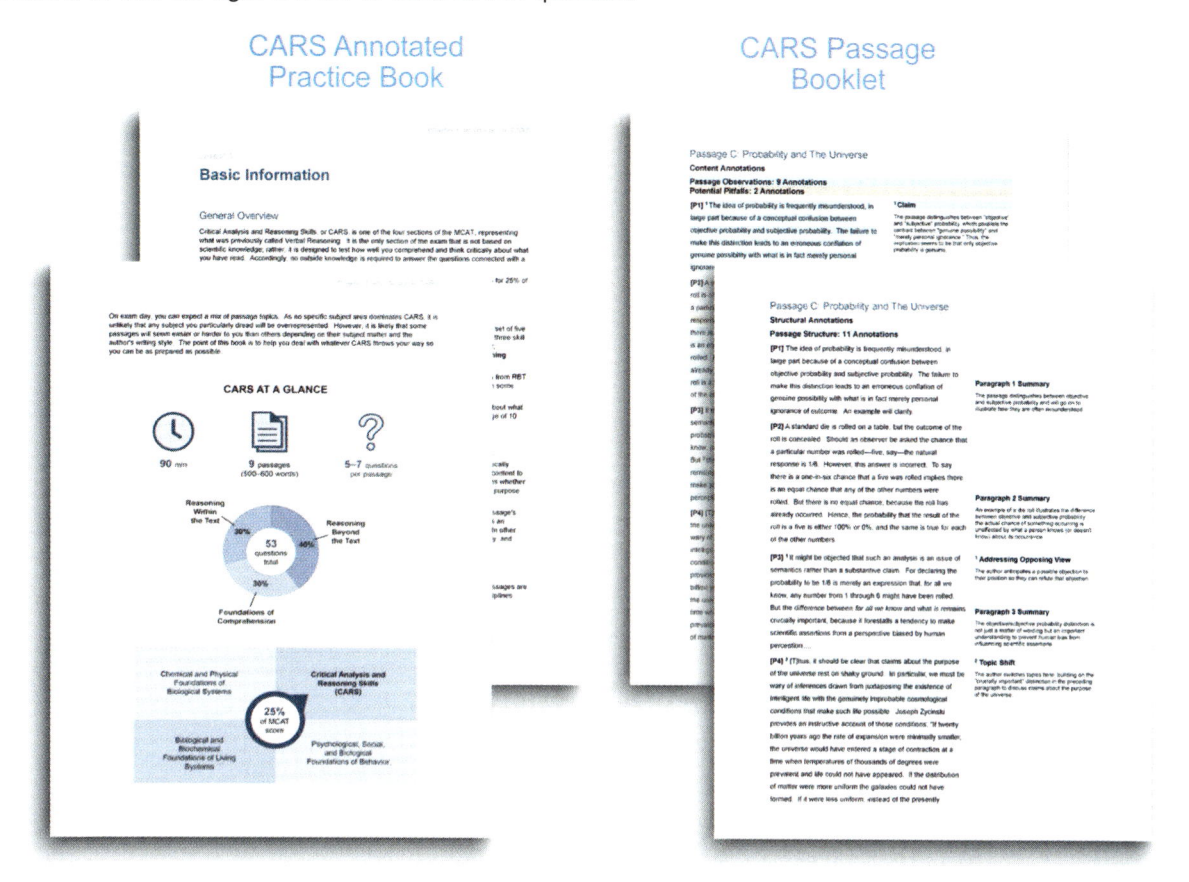

### CARS Annotated Practice Book

### CARS Passage Booklet

# MCAT-Level Exam Practice with the UWorld Qbank

UWorld's MCAT UBooks and Qbank were designed to be used together for a comprehensive review experience. The UWorld Qbank provides an active learning approach to MCAT prep, with thousands of MCAT-level questions that align with each UBook.

The printed UBooks include a prompt at the end of each unit that explains how to access unit practice tests in the MCAT Qbank. In addition, the MCAT UBooks' digital platform enables you to easily create your own unit tests based on each MCAT subject.

To purchase MCAT Qbank access or to begin a free seven-day trial, visit gradschool.uworld.com/mcat.

## Boost Your Score with the #1 MCAT Qbank

Scan for free trial

**Why use the UWorld Qbank?**

- Thousands of high-yield MCAT-level questions
- In-depth, visually engaging answer explanations
- Confidence-building user interface identical to the exam
- Data-driven performance and improvement tracking
- Fully featured mobile app for on-the-go review

# Special Features Integrating Digital UBooks and the UWorld Qbank

The digital MCAT UBooks and the MCAT Qbank come with several integrated features that transform ordinary reading into an interactive study session. These time-saving tools enable you to personalize your MCAT test prep, get the most out of our detailed explanations, save valuable time, and know when you are ready for exam day.

## My Notebook

My Notebook, a personalized note-taking tool, allows you to easily copy and organize content from the UBooks and the Qbank. Simplify your study routine by efficiently recording the MCAT content you will encounter in the exam, and streamline your review process by seamlessly retrieving high-yield concepts to boost your study performance—in less time.

## Digital Flashcards

Our unique flashcard feature makes it easy for students to copy definitions and images from the MCAT UBooks and Qbank into digital flashcards. Each card makes use of spaced repetition, a research-supported learning methodology that improves information retention and recall. Based on how you rate your understanding of flashcard content, our algorithm will display the card more or less frequently.

## Fully Featured Mobile App

Study for your MCAT exams anytime, anywhere, with our industry-leading mobile app that provides complete access to your MCAT prep materials and that syncs seamlessly across all devices. With the UWorld MCAT app, you can catch up on reading, flip through flashcards between classes, or take a practice quiz during lunch to make the most of your downtime and keep MCAT material top of mind.

## Book and Qbank Progress Tracking

Track your progress while using the MCAT UBooks and Qbank, and review MCAT content at your own pace. Our learning tools are enhanced by advanced performance analytics that allow users to assess their preparedness over time. Hone in on specific subjects, foundations, and skills to iron out any weaknesses, and even compare your results with those of your peers.

v

## Table of Contents

### UNIT 1: OVERVIEW OF THE CARS SECTION OF THE MCAT

#### CHAPTER 1: INTRODUCTION TO CARS
Lesson 1.1  Basic Information ................................................................................................... 1

#### CHAPTER 2: THE VALUE OF CARS
Lesson 2.1  Why Do I Have to Study this Stuff, Anyway? ........................................................ 3
Lesson 2.2  Choice of Subject Matter ....................................................................................... 4
Lesson 2.3  Interpreting Information Correctly ......................................................................... 5
Lesson 2.4  Communication Across Backgrounds .................................................................. 6

### UNIT 2: STUDYING CARS

#### CHAPTER 3: HOW TO STUDY FOR CARS
Lesson 3.1  Mindset .................................................................................................................. 7
Lesson 3.2  Practicing Effectively ............................................................................................. 8
Lesson 3.3  Helpful College Courses ....................................................................................... 9

#### CHAPTER 4: APPROACHING CARS PASSAGES AND QUESTIONS
Lesson 4.1  Reading CARS Passages .................................................................................... 11
Lesson 4.2  Types of CARS Passages ................................................................................... 15
Lesson 4.3  Analyzing CARS Questions ................................................................................. 17
Lesson 4.4  Types of CARS Questions ................................................................................... 19

### UNIT 3: UNDERSTANDING CARS SUBSKILLS AND QUESTION TYPES

#### CHAPTER 5: SKILL 1 – FOUNDATIONS OF COMPREHENSION
Lesson 5.1  Subskill 1a. Main Idea or Purpose ...................................................................... 22
Lesson 5.2  Subskill 1b. Meaning of Term .............................................................................. 37
Lesson 5.3  Subskill 1c. Direct Passage Claims .................................................................... 53
Lesson 5.4  Subskill 1d. Implicit Claims or Assumptions ....................................................... 77
Lesson 5.5  Subskill 1e. Identifying Passage Perspectives ................................................... 97
Lesson 5.6  Subskill 1f. Further Implications of Passage Claims ........................................ 117

#### CHAPTER 6: SKILL 2 – REASONING WITHIN THE TEXT
Lesson 6.1  Subskill 2a. Logical Relationships Within Passage .......................................... 133
Lesson 6.2  Subskill 2b. Function of Passage Claim ............................................................ 155
Lesson 6.3  Subskill 2c. Extent of Passage Evidence .......................................................... 171
Lesson 6.4  Subskill 2d. Connecting Claims With Evidence ................................................ 189
Lesson 6.5  Subskill 2e. Determining Passage Perspectives .............................................. 205
Lesson 6.6  Subskill 2f. Drawing Additional Inferences ....................................................... 229

#### CHAPTER 7: SKILL 3 – REASONING BEYOND THE TEXT
Lesson 7.1  Subskill 3a. Exemplar Scenario for Passage Claims ........................................ 245
Lesson 7.2  Subskill 3b. Passage Applications to New Context .......................................... 261
Lesson 7.3  Subskill 3c. New Claim Support or Challenge .................................................. 277
Lesson 7.4  Subskill 3d. External Scenario Support or Challenge ...................................... 299
Lesson 7.5  Subskill 3e. Applying Passage Perspectives .................................................... 325
Lesson 7.6  Subskill 3f. Additional Conclusions From New Information ............................. 341
Lesson 7.7  Subskill 3g. Identifying Analogies ..................................................................... 359

#### CHAPTER 8: FINAL THOUGHTS ON CARS
Lesson 8.1  Bringing It All Together ...................................................................................... 375

Lesson 1.1
# Basic Information

## General Overview

Critical Analysis and Reasoning Skills, or CARS, is one of the four sections of the MCAT, representing what was previously called Verbal Reasoning. It is the only section of the exam that is not based on scientific knowledge; rather, it is designed to test how well you comprehend and think critically about what you have read. Accordingly, no outside knowledge is required to answer the questions connected with a CARS passage.

Like the other sections of the MCAT, the scoring for CARS ranges from 118 to 132 and counts for 25% of your overall MCAT score.

## Format of the CARS Section

The CARS section comprises nine passages of 500-600 words, each of which is followed by a set of five to seven multiple-choice questions, for a grand total of 53 questions. These questions fall into three skill types: **Foundations of Comprehension (CMP)**, or general understanding of passage content; **Reasoning Within the Text (RWT)**, or more critical analysis of passage material; and **Reasoning Beyond the Text (RBT)**, or broader application of principles or ideas from the passage.

On any single exam, roughly 30% of the questions come from CMP, 30% from RWT, and 40% from RBT. All the information you need to answer the questions is contained within the passage, although some RBT questions may also introduce new information in the question itself.

CARS passages tend to be complex, which means you must read carefully and think deeply about what you read. During the exam, you will have 90 minutes to complete the entire section, an average of 10 minutes per passage and question set.

## Purpose of the CARS Section

The primary purpose of the CARS section is to test your reading skills and ability to reason logically based on passage content. CARS also tests whether you can apply what you learn from that content to external situations or in conjunction with new information. In other words, this section evaluates whether you can process complex information and think on your feet. (Chapter 2 further discusses the purpose and value of CARS as related to a career in medicine.)

CARS questions may ask you to identify the main idea of the passage; to consider how the passage's themes or concepts are related; to recognize how the author uses a particular term or supports an argument; or to discern viewpoints and attitudes, draw conclusions, and complete analogies. In other words, you must be able not only to understand but also to analyze, synthesize, interpret, apply, and extrapolate from the material provided in the passage.

## Passage Subject Areas

All CARS passages concern subjects from either the humanities or social sciences. These passages are typically excerpted from articles published in scholarly journals, covering a broad range of disciplines including literature, history, art, music, dance, religion, philosophy, anthropology, archaeology, geography, economics, political science, and psychology, among others.

On exam day, you can expect a mix of passage topics. As no specific subject area dominates CARS, it is unlikely that any subject you particularly dread will be overrepresented. However, it is likely that some passages will seem easier or harder to you than others depending on their subject matter and the author's writing style. The point of this book is to help you deal with whatever CARS throws your way so you can be as prepared as possible.

## CARS AT A GLANCE

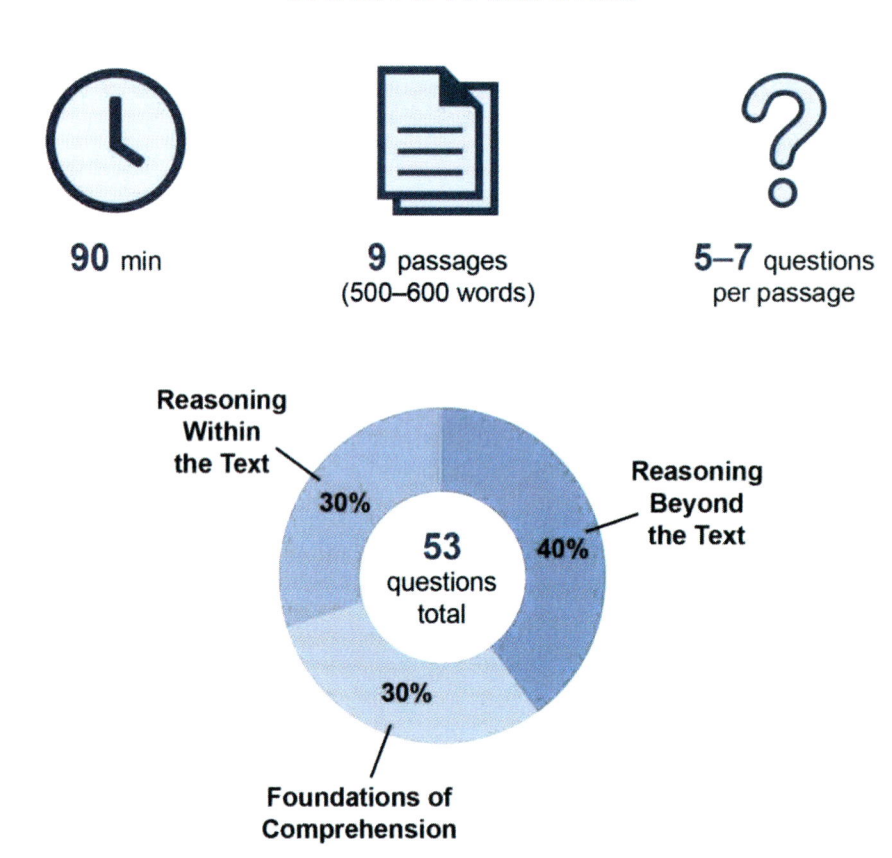

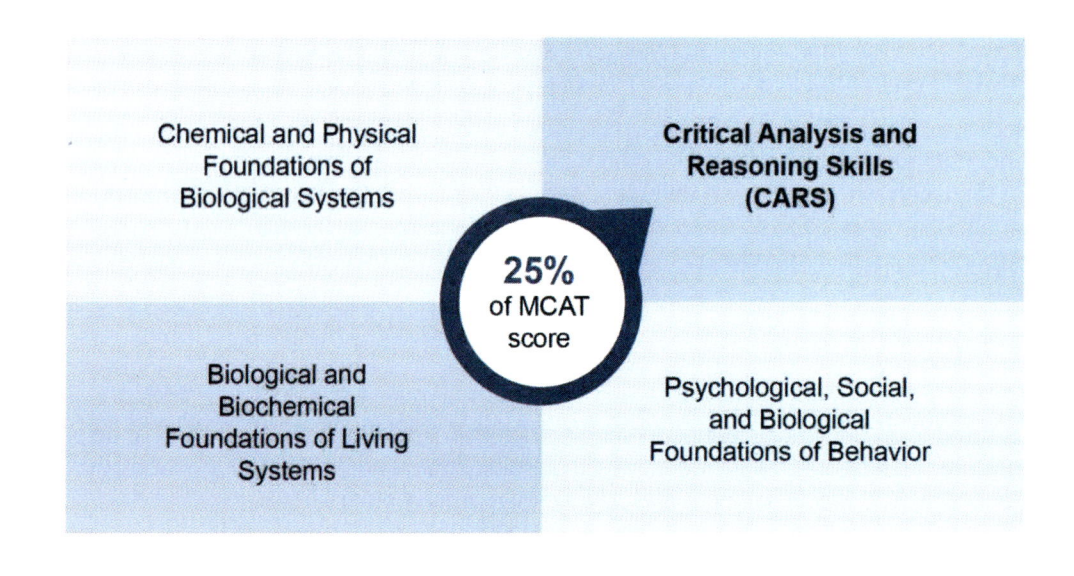

**90** min

**9** passages
(500–600 words)

**5–7** questions per passage

**53** questions total
- Reasoning Within the Text: 30%
- Reasoning Beyond the Text: 40%
- Foundations of Comprehension: 30%

**25% of MCAT score**

- Chemical and Physical Foundations of Biological Systems
- Critical Analysis and Reasoning Skills (CARS)
- Biological and Biochemical Foundations of Living Systems
- Psychological, Social, and Biological Foundations of Behavior

Lesson 2.1
# Why Do I Have to Study This Stuff, Anyway?

At first, the inclusion of the CARS section on the MCAT might seem strange. While it is easy to see how other sections of the exam relate to studying medicine, you may have wondered what relevance CARS really has to your medical education or future career as a doctor. This feeling is only amplified if you find CARS challenging; the more difficult it seems, the more you may want to dismiss it as pointless.

As tempting as that viewpoint can be, however, the type of thinking CARS requires is part of the MCAT for a reason. The exam is designed to test not only your knowledge of science, but also your ability to approach and analyze information logically. Specifically, you might think of CARS as testing you in the following ways:

- Can you draw a reasonable conclusion based solely on the available information?
- Can you analyze a viewpoint as it is stated, without filtering it through your own preconceptions and biases?
- Can you not only understand a claim, but determine what implications would follow if it were true?
- Can you recognize the evidence upon which a statement or argument most depends?
- Can you distinguish legitimate inferences from unjustified assumptions?

Although these skills are tested using material mostly unrelated to medicine, they are widely applicable to your future career. Thus, the CARS section reflects the need for physicians to demonstrate proficiency in a variety of critical reasoning skills.

Lesson 2.2
# Choice of Subject Matter

Most CARS passages address topics from outside the sciences. However, this fact is surprising only if we assume that learning about those *topics* is the point. Consider the names of the four MCAT sections:

- Chemical and Physical Foundations of Biological Systems
- **Critical Analysis and Reasoning Skills**
- Biological and Biochemical Foundations of Living Systems
- Psychological, Social, and Biological Foundations of Behavior

The names of the other three sections refer to subject areas in which you are expected to demonstrate knowledge. By contrast, the name of the CARS section makes no reference to any areas of knowledge; instead, it refers to *reasoning*. So when you start having thoughts like "How does learning about Impressionist painting help me become a better doctor?" the answer is: it doesn't—and it isn't meant to. What does help you become a better doctor is developing critical thinking skills that can be applied regardless of subject matter. That type of thinking—not Impressionist painting, modern economic theory, or whatever—is the true subject area of CARS.

### Skills, Content, and Variables

While it makes sense that critical reasoning skills are important, you might still wonder why these skills are tested using topics in the humanities and social sciences. Couldn't you be given CARS-like questions on content that pre-med students typically study? Perhaps, but doing so would introduce the possibility that questions could be answered using background knowledge instead of solely through critical thinking.

With this fact in mind, it's easier to see why CARS is based on subjects that are less familiar to pre-med students. **Just as research studies are designed to isolate variables, CARS is designed to eliminate the influence of subject knowledge on its questions.** If students do not have content background to rely on, that potential confounding variable is removed, allowing the CARS section to test students' comprehension and reasoning skills exclusively.

Therefore, the fact that CARS usually concerns topics that seem irrelevant to medicine is deliberate: as a software developer might say, "It's not a bug, it's a feature." The point of CARS is to test your ability to reason, and the unfamiliar, sometimes complex content used for the passages helps provide a level playing field on which to test it.

Lesson 2.3
# Interpreting Information Correctly

CARS emphasizes the ability to correctly interpret information. Since there is no pre-existing body of knowledge that you are expected to have absorbed, answering CARS questions is always a matter of accurately recognizing and comprehending passage claims, and reasoning effectively about them.

However, as Lesson 2.1 suggests, there are common ways in which this process can go wrong. Obviously, you might come to an incorrect conclusion if you misunderstand the passage. Beyond that possibility, however, are several related types of error. In CARS, choosing the wrong answer is often the result of:

- overlooking or ignoring relevant statements;
- drawing a conclusion from insufficient evidence;
- following your preconceived ideas instead of focusing on passage claims.

These kinds of flawed interpretations are all based on a faulty approach to the passage and its questions; one that fails to take proper account of the information provided. For instance, if a passage considers a particular idea, but later offers evidence that contradicts it, it would be easy to draw a conclusion too hastily.

Similarly, a diagnosis is also a type of interpretation: a conclusion based on the patient's description of their symptoms and the doctor's own observations. Accordingly, that interpretive process is subject to the same kinds of mistakes, and the same types of critical thinking skills are needed to avoid them.

Lesson 2.4
# Communication Across Backgrounds

One source of frustration in CARS is that sometimes it can feel like the passages are purposely written to be hard to understand. The authors use unusual terms and odd sentence constructions, adding unnecessary complexity to whatever is being discussed. Surely the passages would be much easier if they were written in a way that people actually talk.

However, this sentiment assumes that what counts as natural speech patterns is consistent across people and groups, which is simply not the case. For instance, most of your patients will not come to you with clearly organized descriptions of their problems using standard and unambiguous terminology. Instead, they will convey their ideas in language that makes sense to them, based on the individual backgrounds, cultures, and experiences that affect how each person perceives the world. You must be able to accurately respond to this communication all the same.

### Why Can't They Write Like Regular People?

It's easy to think that some CARS authors write in a way that no regular person would. However, remember that for most CARS passages, **you were not the intended audience**. The works from which these passages are drawn were not written for the purpose of *testing pre-med students* or *providing material for the MCAT*. Typically, they were written for scholars in a specific discipline, using common conventions, techniques, and phrasings that would seem entirely normal to the people expected to read them.

Keeping this fact in mind can help some passages seem less frustrating. In addition, understanding why and for whom a passage was written is important for answering certain questions, particularly those that relate to the author's purpose in writing the passage.

Just as many CARS questions require you to determine an author's view, your success as a doctor will depend significantly on your ability to comprehend the perspectives conveyed by patients and other medical personnel—even when those ideas are expressed in a way that is difficult to understand.

Lesson 3.1
# Mindset

For many students, CARS can seem the most intractable of the MCAT sections, the one most difficult to prepare for and master. In fact, it might even seem as if CARS is the chief obstacle coming between you and your desired MCAT score, causing you to resent the time you spend studying for it.

Chapter 2 explores the purpose behind CARS and the reasons for its inclusion on the MCAT. Whether or not you have come to feel that studying CARS is useful apart from the score you hope to achieve, you can only enhance your preparation if you cultivate a mindset conducive to approaching this section of the exam.

What do we mean by "mindset"? Mindset can be thought of as the overall orientation or attitude you bring to a task, goal, or project. Your mindset comprises the assumptions you have formed—sometimes even unconsciously—about such a project. Furthermore, motivation and behavior tend to flow from the mindset you adopt toward something, thus determining the energy you are willing and able to invest in it.

Consider your mindset regarding CARS. How do you think it affects your motivation and productivity? Do you believe you can develop skills that will help you succeed on this section of the exam? This book aims to remove some of the angst surrounding CARS and to give you the tools for reaching your target score.

### Would You Agree to These Terms?

Suppose someone made you the following offer:

"You can fulfill your dream of being accepted to medical school and becoming a doctor. However, there is one condition: You must first spend an hour and a half reading essays and correctly answering some questions about them."

Obviously, this description is oversimplified. Nevertheless, this is ultimately the situation you face when it comes to the CARS section of the MCAT. By maintaining the perspective that success in CARS is simply one component of achieving your goal, the time and preparation required to achieve that success may feel like less of a burden. Likewise, by adopting the mindset that CARS is a temporary obstacle, you may reduce frustrations as you study and find it easier to overcome.

Lesson 3.2
# Practicing Effectively

Whether you are studying CARS for the first time or have done some preparation before, you probably have a sense of the challenge it represents and may already be feeling the pressure to perform. However, while the test is meant to challenge you, it is not insurmountable. The more familiar you become with CARS, the more you will be able to approach it successfully.

## Plans and Goals

To achieve your desired score, you not only need to be committed but to have a plan. As a rule, you will want to allow yourself at least a few months to prepare for CARS—ideally, integrated within your overall MCAT preparation and not just as one part of it. To help keep yourself on track, formulate a plan that breaks down your overarching goal into specific, smaller goals or intermediate steps.

It is important to think about exactly what such goals will entail. For example, you might aim to complete a certain number of CARS passages and questions per practice session. Alternatively, some of your goals for a given time period might be performance based; for instance, achieving a certain average score for the passages you work on.

Whatever your intermediate steps, you should make practicing CARS into a habit, either incorporating it into your daily routine or designating specific days and times of the week for it. Keep in mind that in the long run, regular and consistent practice sessions are more likely to produce the results you desire than more intense but less frequent sessions.

## Timing

When you first approach CARS, it may be helpful to begin by reading passages and answering questions without paying attention to the clock so you can focus on accuracy rather than speed. Next, you might start timing yourself for some of the passages you work on, gradually transitioning to all of your practice being timed. Eventually, you will want to practice completing nine passage and question sets within a 90-minute period, as allotted on the exam. In your final stage of preparation, you will benefit from taking full-length practice exams that simulate the MCAT in its entirety.

To get the most from your CARS practice sessions, try to work where you will not be disturbed and can control external distractions. If your schedule permits, you should practice at those times of day when you feel the most energetic and alert. However, to help simulate test day conditions, you may also want to try doing some of your preparation during the time of day at which you will be taking the exam.

## Going the Distance

While committing to a plan is important, it is equally important not to overwhelm yourself or undermine what you hope to accomplish. Accordingly, you should set goals that are realistic—that is, they should be challenging but doable. Also, remember to cut yourself some slack from time to time. You will miss questions, and that's okay. The crucial thing is to try to learn from your mistakes and take away something that you can apply later. It's even okay to miss a practice session now and then as long as your overall commitment remains strong.

Keep in mind, too, that progress and improvement in CARS will occur over time, not overnight. The road to mastery will have inevitable setbacks, but setbacks can be overcome, and failures often turn out to be opportunities for determining how to do better.

Lesson 3.3
# Helpful College Courses

As previous lessons discuss, there is no specific subject matter that you are expected to know for CARS. Thus, an additional difficulty in studying for CARS is that it does not appear to be associated with any college course material. For other parts of the MCAT you can take classes that are obviously relevant; for instance, there is little doubt that taking Cell Biology will help prepare you for the Biological and Biochemical Foundations of Living Systems section of the exam. With CARS, on the other hand, it may seem like your only option is just to plow your way through lots of passage and question sets.

However, there actually are college courses that can improve your performance in Critical Analysis and Reasoning Skills. In fact, critical reasoning—and, more generally, logic—is considered a sub-discipline of Philosophy. Furthermore, philosophy classes typically involve reading dense material about complex ideas, analyzing authors' claims, and evaluating arguments. Does this sound familiar?

| Components of Success in CARS | Skills Emphasized in Philosophy Courses |
|---|---|
| • Understand passages of varying complexity and writing style, dealing with many subjects | • Read complicated writings on numerous, often abstract, topics |
| • Discern the author's particular viewpoint and distinguish it from other perspectives | • Analyze authors' views and identify the evidence offered to support them |
| • Evaluate information as presented without referring to one's own beliefs or background knowledge | • Recognize the objective grounds that make a belief justified or unjustified |
| • Identify arguments and the kinds of claims that would support or challenge them | • Construct arguments and evaluate them for strength and logical consistency |

As you can see, the types of skills emphasized in philosophy courses closely reflect those needed for success in CARS. If you can make time in your schedule, you will probably find that taking a philosophy class will improve your CARS performance.

Based on the types of analysis and reasoning that the CARS section requires, we would recommend an Introduction to Philosophy course as likely to be the most helpful. You could also take Critical Thinking or another introductory level philosophy course. Note that while a Logic class could be useful as well, these sometimes focus on more formal systems of reasoning that do not as directly match up with CARS. Finally, other humanities courses, such as those devoted to textual analysis or rhetoric, might also be valuable for your CARS study. Overall, however, a philosophy class will probably provide you with the greatest benefit.

Lesson 4.1

# Reading CARS Passages

## The CARS Difference

As the name Critical Analysis and Reasoning Skills suggests, the CARS section involves *skills you develop through practice* rather than facts you accumulate through study. Therefore, you will have to prepare differently for this section of the exam than for the others.

You might ask, "But isn't CARS mainly a matter of reading?" However, the answer is more complicated than it seems. While CARS does test reading comprehension, many of the passages require you to think about their content in ways with which you might not be familiar.

Scientific texts are typically straightforward and even formulaic in their presentation of material. By contrast, CARS passages are drawn from multiple disciplines, each with its own conventions. Consequently, these passages may not be written in the most direct, linear fashion. If you view yourself primarily as a scientist and fact-oriented individual, both the subject matter and style of some CARS passages could prove off-putting.

To illustrate some of the reading challenges that CARS presents, let's compare the introductory paragraphs of an MCAT-style CARS passage with an MCAT-style biology passage.

| CARS Passage Excerpt | Biology Passage Excerpt |
| --- | --- |
| In August of 1881, the German philosopher Friedrich Nietzsche was walking beside a Swiss Alpine lake when a thought that he would alternately extol and execrate as "the thought of all thoughts"—one that would beleaguer the rest of his life and work—flashed through his mind. The thought first appears in Nietzsche's 1882 collection of aphorisms, *The Gay Science*, as a hypothetical question: "What if some day or night a demon were to steal after you into your loneliest loneliness and say to you: 'This life as you now live it and have lived it, you will have to live once more and innumerable times more; and there will be nothing new in it, but every pain and every joy and every thought and sigh and everything unutterably small or great in your life will have to return to you, all in the same succession and sequence....?'" A curious intertwining of infinity, repetition, and fate, known as the Eternal Recurrence of the Same, this idea would itself recur in the philosopher's writing, particularly in his self-proclaimed masterwork and quasi-hermetic text *Thus Spake Zarathustra*. | In the 1800s, cell theory studies included research into cell reproduction and major cell structures. Specifically, an early hypothesis regarding animal cell reproduction predicted that all cellular components arose from a fluid substance called cytoblastema. It was thought that during reproduction, a portion of an existing cell's cytoblastema would crystallize into a dense nucleolus, followed by the deposition of additional crystallized cytoblastema to form a new nucleus. Researchers believed that a new, separate cytoblastema layer would then thicken around this new nucleus, ultimately forming a daughter cell. |

| CARS Excerpt Features | Biology Excerpt Features |
|---|---|
| Discusses a philosophical theory with some historical and biographical context. | Outlines a scientific theory retrospectively. |
| Uses varied, elevated, and metaphorical vocabulary.<br>Examples:<br>• "extol"<br>• "execrate"<br>• "beleaguer"<br>• "aphorisms"<br>• "quasi-hermetic" | Uses standard terminology that is repeated.<br>Examples:<br>• "cytoblastema"<br>• "reproduction"<br>• "nucleus" |
| Ornate, sometimes unique phrasing; conveys a sense of gravity.<br>Example:<br>• "…a curious intertwining of infinity, repetition, and fate known as the Eternal Recurrence of the Same, this idea would itself recur in the philosopher's writing, particularly in his self-proclaimed masterwork and quasi-hermetic text *Thus Spake Zarathustra*." | Straightforward; recites a process. Creates a more neutral atmosphere.<br>Examples:<br>• "It was thought that…"<br>• "Researchers believed that…" |
| Tends to be wordy rather than concise; includes an extended quotation. Phrasing and stylistic features contribute to the overall length of the paragraph. | Tends to be concise rather than wordy; longer sentences result from the compounding of information rather than from stylistic features. |
| Largely subjective, meant to stir emotion and curiosity. | Largely objective and unemotional. |

This comparison offers some insight into the differences between the types of passages you will encounter on the CARS section and those you will read for the science sections. It may also give you a sense of the nuances that CARS passages may contain. If you have practiced CARS before, did you ever feel like you were being swamped by metaphors, rhetorical devices, and unfamiliar jargon? Did sentences seem convoluted or unnecessarily long-winded? Perhaps the subject matter appeared needlessly dense, deliberately designed to bore you, or even cryptic.

Fortunately, the more familiar you become with the various features of CARS passages, the more adeptly you can navigate their twists and turns. These passages often contain subtleties that require you to pay attention to multiple things at once. The key to managing all these demands is to adopt a critical attitude of **continually questioning the passage as you read**. You should always be asking yourself: "What is the passage trying to communicate to me, and why should I believe it?" Answering this question involves looking carefully at different aspects of the passage, with special attention to **Style**, **Tone**, **Viewpoint**, and **Main Idea**.

## Style

The way an author presents a text can greatly influence how it is perceived. Thus, a passage may express its message partly through the author's choice of words and the way sentences are structured. To take a simple example, the use of elevated vocabulary might make a passage seem lofty or authoritative; alternatively, the use of metaphors or other figures of speech might give it a poetic character. It is important to note that such stylistic features often reinforce or mirror the overall point the author is trying to get across.

## Tone

Another aspect of passage presentation is tone. Through the tone they use, authors can impact how readers should think or feel about the material being discussed. Writers may purposely use language to arouse emotions, shape perceptions, and contribute to evoking a particular state of mind in the reader. For instance, abrupt phrases and sharp words can transmit an impatient or dismissive tone, or terms of insincere praise could be used to create a mood of irony or sarcasm.

Notice that science passages also have a tone, but typically one that is more flat or neutral. As an example, let's think back to our passage comparison for a moment. The CARS passage portrays Nietzsche's thought as "curious," suggesting that the topic is intriguing. By contrast, it would be hard to imagine the science passage being phrased in a way that conveyed any sort of bias—for instance, if it referred to "a *ridiculous* early hypothesis" or declared that "*fortunately*, researchers believed...."

Therefore, while often subtle, tone is an important element of any text and can "speak" loudly. As you read through CARS passages, try to discern the general mood, asking yourself, "How am I being asked to think or feel about this topic?"

## Viewpoint

Closely connected to the tone of a passage is the perspective or point of view from which it is written. The author's attitude or opinion often comes through in the way they portray their subject. Do they see it as a problem to be solved or maybe as a phenomenon to be explored? Are they offering an even-handed analysis, or do they favor one side of a debate?

Keep in mind that a passage may include perspectives other than the author's. Looking once again at our comparison between the CARS and Biology passage excerpts, we can see how the CARS passage presents *both* Nietzsche's and the author's perspectives on the "Eternal Recurrence." Specifically, the passage mentions that Nietzsche would "alternately extol and execrate" what he called "'the thought of all thoughts,'" then goes on to depict the author's own view of that thought as a "curious intertwining" of concepts.

Thus, recognizing perspectives is another component essential to understanding a passage, and some CARS questions may even ask you directly about viewpoints or attitudes. Lessons 5.5, 6.5, and 7.5 discuss these types of questions in depth.

## Main Idea

When reading CARS passages, you may find yourself wanting to know right away what they're all about, or wishing they would just get to the point as quickly as possible. Try to make that impatience work to your advantage: As you read, ask yourself, why did the author write this? What is the message they are trying to convey? In other words, you are in search of the thesis—the central, organizing statement that gives the passage its focus.

The thesis of a passage may be expressed in one sentence and is often, but not always, found at the end of the first paragraph. Sometimes, however, the thesis is less obvious, perhaps spread over more than one sentence or merely implied. Beware that a tricky passage can act as a bait-and-switch: it might begin by discussing one topic before taking up another as its true focus. Lesson 5.1 explores how to approach main idea questions in greater detail.

## Putting It into Practice

While this may seem like a lot to pay attention to, you will probably find that doing so becomes easier and more natural over time. As you practice with the passages in Chapters 5, 6, and 7, you can also refer to the CARS Passage Booklet, which provides structural and content-based annotations for each passage. These annotations illustrate the kinds of features that are important to notice for reading and analyzing passages effectively.

Lesson 4.2
# Types of CARS Passages

## Many Subjects, Few Types

As discussed in Lesson 1.1, CARS passages are drawn from diverse disciplines in the humanities and social sciences. Thus, the topics on your exam could come from as many as nine different subject areas out of even more potential choices. Given these possibilities, you might worry that you have to prepare for many different kinds of passages.

But despite the wide variation in subjects, there are actually only a few *types* of CARS passages. To begin with, almost all passages can be categorized as basically either **Argumentative** or **Descriptive**.

| Argumentative | Descriptive |
|---|---|
| Intended to **convince the reader** of something<br><br>Examples:<br><br>• a philosopher presents **reasons why** we should believe one thing instead of another<br>• an anthropologist **defends** a theory about a past civilization<br>• an art expert **makes a case** for how a painting should be interpreted | Intended to **provide information** about a topic<br><br>Examples:<br><br>• a historian **recounts** the circumstances leading to a significant event<br>• a sociologist **explains** the similarities between two cultures<br>• an economist **illustrates** a relationship between different consumer behaviors |

Argumentative passages are just what they sound like: works whose main purpose is to *convince the reader* of a claim or viewpoint. Regardless of the topic, such passages use evidence and reasoning to persuade readers to accept a particular idea.

Descriptive passages, on the other hand, seek primarily to *provide information* about the subject being discussed. Whatever that subject is, these passages are designed to convey facts, usually in an objective manner.

As you might guess, *the two types are not mutually exclusive*: for instance, a passage might first describe a set of known facts, then provide an argument about a disputed point. Still, for the most part every passage will fit into one of these two categories.

## Different Topics, Similar Structures

Although CARS passages cover many different topics, there are similarities in how the passages are structured.

For example, consider two passages in the CARS Passage Booklet: Passage C, *Probability and The Universe*, and Passage G, *Food Costs and Disease*. As you read these passages and review their structural annotations, keep the following questions in mind:

- What is the author's purpose in writing the passage?
- How does the organization of the passage relate to that purpose?

Now let's compare them. If you are like many students, you may have preferred *Food Costs and Disease*. Not only is it a Population Health passage with content based on facts and empirical studies, its topic of disease prevention is relevant to a career in medicine. By contrast, *Probability and The Universe* is a Philosophy passage dealing with abstract conceptual claims. It starts with an odd discussion about probability, then moves into a complicated argument about the purpose of the universe. Nevertheless, these two argumentative passages are organized in much the same way.

| Passage Structure: *Probability and The Universe* | Passage Structure: *Food Costs and Disease* |
| --- | --- |
| • The passage begins by explaining subjective versus objective probability **(P1)**, then illustrates this distinction with an example **(P2)**. | • The passage begins by explaining the relationship between food price adjustments and reduced disease **(P1)**, then illustrates this relationship with data **(P2)**. |
| • Building on the previous example, the passage introduces a new topic: the effects of human bias on scientific understanding **(P3)**. The author specifically suggests that we cannot conclude there is a purpose to the universe, despite how improbable its existence may seem **(P4)**. | • Building on the data provided, the passage switches to a new topic: objections to other effects of adjusting food prices **(P3)**. The author first responds to an objection about the financial burden of such adjustments **(P4)**. |
| • The author more directly argues that there may be no purpose to the universe, referencing the example from Paragraph 2 as an illustration **(P5)**. The author ends by reinforcing this conclusion, tying it back to the point about probability from the beginning of the passage **(P6)**. | • The author responds to another objection, that food price adjustments could violate people's autonomy **(P5)**. The author ends by arguing that price adjustments either do not violate autonomy or are preferable to the alternative even if they do **(P6)**. |

In addition to discussing different topics, one passage is theoretical in nature while the other is based more on concrete data. As you can see, however, their structures are largely parallel. First, each uses an initial subject to introduce another one that is treated more extensively. Then, each offers an argument that addresses competing views and establishes a conclusion about the new subject.

As you start to recognize structural patterns in passages, studying for CARS becomes a much more straightforward process. Even more significantly, CARS questions also fall into patterns, which later lessons explore in depth.

Lesson 4.3
# Analyzing CARS Questions

## Slower Is Faster

As mentioned in Lesson 1.1, the CARS section of the MCAT comprises 9 passage and question sets that you must complete in 90 minutes, giving you an average of 10 minutes per set. Although it may be tempting to rush through the passage to get to the questions sooner, there are good reasons for doing just the opposite.

First, let's break down the time factor. Most CARS passages are between 500 and 600 words, and the average adult reading speed is 250 words per minute. Thus, at an average reading pace, it would take between two and three minutes to get through each passage. However, some passages are more complicated than others, and you may want to re-read certain parts to make sure you understand them. So, let's suppose you took four minutes to read a passage thoroughly. With an average of six questions (five to seven per passage), that would still leave you a full minute to answer each of them.

Second, many CARS questions are based on connecting multiple pieces of information or understanding the passage as a whole. If you only skim the passage, you will likely lack this fuller understanding, making those questions more difficult than they need to be and forcing you to re-read the passage anyway. By focusing on comprehending what you read in the first place, *you will likely spend less time overall*, even if you sometimes refer back to review or confirm particular details.

Therefore, we recommend that the first step in analyzing questions should be: **Read each passage at a natural pace, focusing on comprehension**. As you do, take note of the passage title, which often gives a sense of the overall topic and can even help in answering certain questions.

### Highlighting

Some students find it useful to highlight the passage as they read, while others do not. Consider trying both approaches to see what works best for you. Keep in mind, however, that certain types of highlighting tend to be more helpful than others. As a general rule, **highlighting should make it easier to refer back to details when needed**. For example, highlighting *transition words* ("however"; "but"; "on the other hand"), *viewpoint indicators* (see Lessons 5.5, 6.5, and 7.5), and statements that *elaborate on the main idea* will likely assist you in answering questions more quickly and easily.

## Understanding the Question

One of the easiest ways to answer a question incorrectly is to misinterpret it. For example, suppose a question asked:

> "Which of the following claims would most undermine the author's argument for providing larger pensions to government employees?"

An answer choice that mentions the high cost of such pensions would seem to count against the author's proposal. However, if the author's argument never discusses the cost of the pensions but is instead based entirely on other factors, then *that particular argument* would not be undermined. While the answer choice might make sense from a general standpoint, it would not address what the question actually asked.

If a question is long or confusingly worded, take an extra moment to make sure you understand it, and perhaps rephrase it in more natural terms. It is especially important to take note of negative words like NOT, LEAST, or EXCEPT, which reverse what would otherwise be the correct answer. You may even want to highlight such terms to keep them in the front of your mind.

## Ignoring Distractions

While it seems natural to read the answer choices along with the question, we recommend a different approach: **Whenever possible, formulate an answer to the question *before* looking at the answer choices**.

By doing so, you are focusing on what the right answer should be while avoiding becoming distracted by the incorrect answer choices. (In fact, test writers often refer to the incorrect choices as "distractors.") Thus, your starting point is not four potential answers, three of which you need to rule out; instead, your starting point is your idea of what the right answer is. Then, when you do review the answer choices, you are simply looking for the one that best matches what you have already determined is correct. This method will both save you time and keep you from being swayed by answer choices designed to trick you.

> **What If There Were No Answer Choices?**
>
> Imagine that the CARS section was not multiple choice, but instead simply required you to write out an answer for each question. How would you come up with these answers? You would have to read the passage and determine logically what the best response was—just like you need to do already.
>
> Accordingly, by ignoring the answer choices at first, you are essentially taking the wrong ones out of the equation, making it easier to avoid talking yourself into them.

We can sum up our general approach to answering questions as follows:

**Read each passage at a pace that seems natural to you**.
- Don't rush; focus on understanding.
- Highlight terms that help you refer back to details when needed.

**Pay attention to what questions are really asking**.
- Rephrase questions if necessary.
- Take special note of negative words like NOT, LEAST, and EXCEPT.

**Whenever possible, formulate an answer *before* looking at the answer choices**.
- Find the answer choice that best matches your idea of the correct answer.
- Don't be afraid to refer back to the passage for confirmation.

Finally, the way that you formulate your answers will depend in part on the *type* of questions you are dealing with. The remainder of this chapter begins discussing those types, while Chapters 5, 6, and 7 go into much greater detail.

Lesson 4.4
# Types of CARS Questions

The MCAT© divides Critical Analysis and Reasoning Skills into three overarching skills:
- **Foundations of Comprehension (CMP)**
- **Reasoning Within the Text (RWT)**
- **Reasoning Beyond the Text (RBT)**

Broadly speaking, Foundations of Comprehension questions test your basic understanding of the passage; Reasoning Within the Text questions concern relationships between passage claims; and Reasoning Beyond the Text questions introduce outside information that you must consider in conjunction with the passage.

However, the more you practice CARS, the more it becomes clear that a variety of question types are included within each of the three skills. At UWorld, we have broken those skills down into a total of 19 *subskills*, smaller categories that more precisely reflect the different kinds of questions that are asked in CARS.

| Skill 1: Foundations of Comprehension ||
|---|---|
| Subskill | Student Objective |
| 1a. Main Idea or Purpose | Identify the central theme or purpose of the passage |
| 1b. Meaning of Term | Determine the meaning of a term as used in the passage |
| 1c. Direct Passage Claims | Identify claims made directly in the passage |
| 1d. Implicit Claims or Assumptions | Identify implicit claims or unstated assumptions in the passage |
| 1e. Identifying Passage Perspectives | Identify perspectives or positions presented in the passage |
| 1f. Further Implications of Passage Claims | Recognize further implications of claims stated in the passage |

## Skill 2: Reasoning Within the Text

| Subskill | Student Objective |
|---|---|
| 2a. Logical Relationships Within Passage | Identify logical relationships between passage claims |
| 2b. Function of Passage Claim | Determine what function a particular claim serves in the passage |
| 2c. Extent of Passage Evidence | Determine the extent to which the passage provides evidence for its claims |
| 2d. Connecting Claims With Evidence | Connect passage claims with supporting passage evidence |
| 2e. Determining Passage Perspectives | Determine the perspective of the author or another source in the passage |
| 2f. Drawing Additional Inferences | Draw additional inferences based on passage information |

## Skill 3: Reasoning Beyond the Text

| Subskill | Student Objective |
|---|---|
| 3a. Exemplar Scenario for Passage Claims | Identify which scenario most exemplifies or logically follows from passage claims |
| 3b. Passage Applications to New Context | Apply passage information to a new situation or context |
| 3c. New Claim Support or Challenge | Evaluate how a new claim supports or challenges passage information |
| 3d. External Scenario Support or Challenge | Determine which external scenario supports or challenges passage information |
| 3e. Applying Passage Perspectives | Apply the perspective of the author or another source in the passage to a new situation or claim |
| 3f. Additional Conclusions From New Information | Use new information to draw additional conclusions |
| 3g. Identifying Analogies | Identify analogies or similarities between passage ideas and ideas found outside the passage |

As you look over this list of subskills, you may notice parallels between some of them.  For example, subskills **1e**, **2e**, and **3e** all relate to distinguishing perspectives in a passage, while subskills **2c** and **2d** represent two different types of RWT questions about passage evidence.  Similarly, subskills **3c** and **3d** correspond to variations in RBT questions about supporting or challenging passage claims.  These and other parallels are further discussed in the following chapters.

Most of this book is dedicated to exploring the various subskills in depth, providing methods for each of them along with illustrated examples and step-by-step breakdowns of important concepts.  By understanding the similarities and differences between these subskills, you will start to get a feel for CARS questions at a more granular level, developing a fuller and more accurate sense of the kinds of questions asked and the types of reasoning required to answer them.

Lesson 5.1
# CMP Subskill 1a. Main Idea or Purpose

In this chapter, we discuss the first competency or skill of the CARS section: **Foundations of Comprehension (CMP)**. While all CMP questions involve a primary understanding of passage claims, they can be further broken down into six question types, each represented by a specific subskill. For each subskill, a description is given, followed by a method to successfully approach the questions testing it. Finally, three to five example questions are provided to illustrate the method and allow you to practice.

| Skill 1: Foundations of Comprehension ||
|---|---|
| **Subskill** | **Student Objective** |
| **1a. Main Idea or Purpose** | **Identify the central theme or purpose of the passage** |
| 1b. Meaning of Term | Determine the meaning of a term as used in the passage |
| 1c. Direct Passage Claims | Identify claims made directly in the passage |
| 1d. Implicit Claims or Assumptions | Identify implicit claims or unstated assumptions in the passage |
| 1e. Identifying Passage Perspectives | Identify perspectives or positions presented in the passage |
| 1f. Further Implications of Passage Claims | Recognize further implications of claims stated in the passage |

**1a. Main Idea or Purpose** questions ask you to consider the overall purpose of the passage. As mentioned in Lesson 2.4, that purpose is not "to test pre-meds hoping to become doctors" or "to provide material for the MCAT." Rather, the passages you will encounter on the MCAT were originally written to convey a particular message to a particular audience. Thus, main idea questions ask you to identify that message.

## Identifying as You Go

In many ways, answering **1a. Main Idea or Purpose** questions is a process that should occur constantly as you read. Part of understanding any written work is to understand the point of writing it. So, to read a passage effectively means you should question why it was written and revise your assessment along the way. Ideally, by the time you finish reading, you will have identified the main idea of the passage.

Does this mean that all main idea questions should be easy? Not necessarily—some passages may have a complex topic or begin discussing one subject as a lead-in to another, making their central point more difficult to identify. But even in these cases, the right approach can make picking out the main idea of the passage a more straightforward process.

## Method for Answering 1a. Main Idea or Purpose Questions

### Step 1: Ask Focus Questions

First, a good way to approach the main idea of a passage is to ask yourself one or more of the following **focus questions**:

1. What does the author primarily want people to know from reading the passage?
2. What does the passage seem most concerned with?
3. If I had to summarize the passage in one sentence, what would that sentence be?

These questions are three ways of asking the same thing, so use whichever version feels most natural to you. When asking these questions, keep in mind that the correct answer to a main idea question should be *comprehensive*. In other words, it should reflect the passage as a whole, not just particular details.

### Step 2: Compare Answer Choices with the Focus of the Passage

The answer to your focus question in Step 1 should capture the main idea of the passage in at least a general sense. Accordingly, the next step is to compare the question's answer choices with your own description of the passage's focus. You will most likely find that one answer choice either closely matches this description or represents a more specific version of it.

### Step 3: Consider Your Chosen Answer as a Thesis or Conclusion

Finally, the main idea of any passage **could serve as either its thesis statement or overall conclusion**. In some cases, the introductory or closing paragraphs may explicitly include such a statement. But even when such a statement is not explicitly included in the passage, the main idea still exists. So, the correct answer will describe a statement that the author could have used as their thesis, even if they did not. Therefore, when evaluating your answer choice, the last step is to ask: Would this answer fit as a thesis statement, overall conclusion, or both? If the answer is yes, you have correctly identified the main idea.

## Subskill 1a. Main Idea or Purpose
## (3 Practice Questions)

We now look at three examples from the UWorld Qbank. In each case, the passage and question are first given without commentary, allowing you to practice applying the method yourself. Then, the passage and question are presented again, this time with annotations of the passage (see the CARS Passage Booklet) and a step-by-step explanation of the question using the described method.

## Passage A: The Knights Templar

The seal of the Knights Templar depicts two knights astride a single horse, a visual testament of the order's poverty at its inception in 1119.  Nevertheless, these Knights of the Temple—who swore oaths not only of poverty but of chastity, loyalty, and bravery—would eventually become one of the wealthiest and most powerful organizations in the medieval world.  So far-reaching was their strength and acclaim that their destruction must have seemed as sudden and surprising as it was utter and irrevocable.  The signs of danger could not have been wholly invisible, however, as the Templars' growing influence became perceived as a threat to European rulers.

The first Templars were nine knights who took an oath to defend the Holy Land and any pilgrims who journeyed there after the First Crusade.  Having secured a small benefice from Jerusalem's King Baldwin II, the knights inaugurated their mission at the site of the great Temple of Solomon.  The order quickly attracted widespread admiration as well as many recruits from crusaders and other knights.  Within a year of its founding, the order received a financial endowment from the deeply impressed Count of Anjou, whose example was soon followed by other nobles and monarchs.  As early as 1128, the Templars even gained official papal recognition, and their wealth, holdings, and numbers swelled both in the Holy Land and throughout Europe, especially in France and England.

However, this growing power contained the seeds of the order's downfall.  Although the Templars were generally held in high esteem, the passage of time saw censure and suspicion directed toward them.  The failure of the disastrous siege of Ascalon in 1153 was attributed by some to Templar greed.  Similarly, in 1208 Pope Innocent III condemned the wickedness he believed to exist within their ranks.  Moreover, their increasingly elevated status brought them into conflict with established authorities.  One revealing example occurred in 1252 when, because of the Templars' "many liberties" as well as their "pride and haughtiness," England's King Henry III proposed to curb the order's strength by reclaiming some of its possessions.  The Templars' response was unambiguous: "So long as thou dost exercise justice thou wilt reign; but if thou infringe it, thou wilt cease to be King!"

Ultimately, the impoverished Philip the Fair of France joined forces with Pope Clement V to engineer the Templars' downfall beginning in 1307.  Conspiring to seize the order's wealth, Clement invited the Templar Master, Jacques de Molay, to meet with him on the pretext of organizing a new crusade to retake the Holy Land.  Shortly thereafter, Philip's forces arrested de Molay and his knights on charges the preponderance of which were almost certainly fabricated.  Ranging from the mundane to the unspeakably perverse, the accusations even included an incredible entry citing "every crime and abomination that can be committed."  Suffering tortures nearly as horrific as the acts of which they were accused, many Templars confessed.

Not everyone believed these charges.  Despite Philip's urging, Edward II of England remained convinced that the accusations were false, a view seemingly shared by other rulers.  Nevertheless, Clement ordered Edward to extract confessions, a task the king tried to carry out with some measure of mercy.  By 1313, the Templar Order had been dissolved by papal decree, and many of its members were dead.  The following year, Jacques de Molay was burned at the stake after declaring that the Templars' confessions were lies obtained under torture.  Stories would spread that as he died, he condemned Clement and Philip to join him within a year.  Although it is impossible to say whether he truly called divine vengeance down upon them, within a few months' time both pope and king had gone to their graves.

*The Knights Templar* ©UWorld

Annotations for this passage can be found in the CARS Passage Booklet.

## 1a. Main Idea or Purpose Question 1

The author's main point is that the Knights Templar:

- A. grew to possess power far beyond what might be expected from their humble beginnings.
- B. were actually a more powerful force in Europe than they were in the Holy Land.
- C. possessed great power as an organization, which ironically brought them to ruin.
- D. possessed sufficient power to challenge the rule of monarchs.

See next page for the *strategy-based explanation* of this question.

# Chapter 5: Skill 1 – Foundations of Comprehension

> ## 1a. Main Idea or Purpose Question 1
> ## Strategy-based Explanation
>
> The author's main point is that the Knights Templar:
>
> A. grew to possess power far beyond what might be expected from their humble beginnings.
> B. were actually a more powerful force in Europe than they were in the Holy Land.
> C. possessed great power as an organization, which ironically brought them to ruin.
> D. possessed sufficient power to challenge the rule of monarchs.

As with any **1a. Main Idea or Purpose** question, we are looking for the idea that best describes the information in the passage as a whole. In this case, the correct answer will be a statement about the Knights Templar that succinctly captures the author's major claims about them in the passage. Therefore, we should see this answer reflected in multiple parts of the passage that all express a common theme about these knights.

## Applying the Method

*Passage Excerpt*

**[P1]** The seal of the Knights Templar depicts two knights astride a single horse, a visual testament of the order's poverty at its inception in 1119. Nevertheless, these Knights of the Temple—who swore oaths not only of poverty but of chastity, loyalty, and bravery—would eventually become **one of the wealthiest and most powerful organizations** in the medieval world. **So far-reaching was their strength and acclaim** that **their destruction** must have seemed as sudden and surprising as it was utter and irrevocable. The signs of danger could not have been wholly invisible, however, as **the Templars' growing influence became perceived as a threat to European rulers**.

**[P2]** As early as 1128, the Templars even gained official papal recognition, and **their wealth, holdings, and numbers swelled** both in the Holy Land and throughout Europe, especially in France and England.

**[P3]** However, **this growing power contained the seeds of the order's downfall**. Although the Templars were generally held in high esteem, the passage of time saw censure and suspicion directed toward them. The failure of the disastrous

### Step 1: Ask Focus Questions

Given how the question is worded, we might begin with this version of a focus question: **What does the author primarily want people to know about the Knights Templar?**

First, the passage covers the entire history of the Knights Templar: it begins with **the order's inception in 1119** and ends with it **being dissolved in 1313**. Thus, the author seems concerned with the rise and fall of the Templars over time.

Second, multiple sections of the passage refer to the **great power** the Templars gained and how that power **led to conflict with European rulers**. So, the effects of the Templars' power is a significant concern of the author.

27

siege of Ascalon in 1153 was attributed by some to Templar greed.  Similarly, in 1208 Pope Innocent III condemned the wickedness he believed to exist within their ranks.  Moreover, **their increasingly elevated status brought them into conflict with established authorities**.

[P4] Ultimately, the impoverished **Philip the Fair of France joined forces with Pope Clement V to engineer the Templars' downfall** beginning in 1307.

[P5] By 1313, the Templar Order had been dissolved by papal decree, and many of its members were dead.  The following year, Jacques de Molay was burned at the stake after declaring that the Templars' confessions were lies obtained under torture.

Based on these observations, what the author primarily wants people to know about the Knights Templar would seem to be:

**How the Templars rose to power, clashed with European rulers, and were eventually destroyed**.

### Step 2: Compare Answer Choices with the Focus of the Passage

In answering our focus question, we determined that the author primarily wants people to know **how the Templars rose to power, clashed with European rulers, and were eventually destroyed**.  Turning to the answer choices, we find that **Choice C** is quite similar to this idea: *possessed great power as an organization, which ironically brought them to ruin*.  Although it does not mention the clash with rulers specifically, this answer choice refers to both the Templars' power and the effects of that power in eventually leading to their destruction.

The only other answer we might consider is Choice D, *possessed sufficient power to challenge the rule of monarchs*.  However, while this answer choice does mention the Templars' challenge to European rulers, it does not include the effects of that conflict in eventually leading to the Templars' destruction.  Therefore, **Choice C** is the more comprehensive answer.

### Step 3: Consider Your Chosen Answer as a Thesis or Conclusion

As a final test of this answer, we can ask whether **Choice C** would fit as a thesis statement or overall conclusion to the passage.  Looking at Paragraph 1, we see that it both emphasizes the Templars' great power and links it to their surprising destruction.  Accordingly, the claim that the Templars *possessed great power as an organization, which ironically brought them to ruin* could easily be included in the introduction as a thesis statement.  Therefore, we can be confident that **Choice C** is the correct answer.

For an alternative method of explanation based more specifically on passage evidence, you can view this question in the UWorld Qbank (sold separately).

## Passage B: Jackie in 500 Words

Born in New York in 1929, Jacqueline Bouvier first came into the public eye as the wife of the 35th president of the United States, John F. Kennedy. The president was assassinated in 1963, and by the end of what turned out to be a turbulent decade, Mrs. Kennedy had transformed herself into the enigmatic Jackie O., wife of Greek shipping magnate Aristotle Onassis. Multifaceted and always elusive, the former first lady never ceased to fascinate; however, people had to be satisfied with only glimpses of this fashion icon, culture advocate, historic preservationist, polyglot, equestrienne, and book editor. Indeed, upon her death in 1994, Jacqueline Kennedy Onassis was described as "the most intriguing woman in the world." Often topping lists of the most admired individuals of the second half of the 20th century, this celebrated woman is likely someone many wish they had known. Barring such a possibility, the best way to fully appreciate Jackie's exceptional nature might be to consider the people she wished she had known.

It is unsurprising that this woman who captured the public's imagination for decades distinguished herself from her peers early on. Notably, in 1951, Ms. Bouvier entered a scholarship contest sponsored by *Vogue* and open to young women in their final undergraduate year, the annual Prix de Paris. Among other assignments, applicants were asked to compose a 500-word essay, "People I Wish I Had Known," spotlighting three individuals influential in art, literature, or culture. The future first lady chose an iconoclastic trio from the Victorian era: the French symbolist poet Charles Baudelaire, the Irish wit Oscar Wilde, and the innovative Ballets Russes dance company founder Sergei Diaghilev.

In a brief composition, Jackie provided deep insights into this bohemian threesome of poet, aesthete, and impresario with whom she strongly identified. She concluded that Baudelaire deployed "venom and despair" as "weapons" in his poetry. She idolized Wilde for being able "with the flash of an epigram to bring about what serious reformers had for years been trying to accomplish." Diaghilev she defined as an artist of a different sort, someone who "possessed what is rarer than artistic genius in any one field—the sensitivity to take the best of each man and incorporate it into a masterpiece." As Jackie poignantly observed, such a work is "all the more precious because it lives only in the minds of those who have seen it," dissipating soon after. Furthermore, although these men espoused different disciplines, she discerned that "a common theory runs through their work, a certain concept of the interrelation of the arts." Finally, foreshadowing her self-assumed role in the White House as the nation's unofficial minister of the arts, Jackie paid homage with her vision: "If I could be a sort of Overall Art Director of the Twentieth Century, watching everything from a chair hanging in space, it is their theories of art that I would apply to my period."

The contest committee judged Jackie's essay to have exhibited a profound appreciation for the arts combined with a truly outstanding level of intellectual maturity and originality of thought. Similarly, biographer Donald Spoto deemed Jackie "remarkably unorthodox," not unlike the men about whom she wrote in her unusual composition, which he pronounced "a masterpiece of perceptive improvisation." Thus, from a pool of 1,279 applicants representing 224 colleges, Jacqueline Bouvier was declared the winner.

Although Ms. Bouvier went on to decline the prestigious award, which would have involved living and working in Paris, she never gave up her dream of being the century's art director. As first lady, she tirelessly promoted the arts and culture. Today, the John F. Kennedy Center for the Performing Arts in Washington, DC, is a legacy of Jackie's vision.

*Jackie in 500 Words* ©UWorld

Annotations for this passage can be found in the CARS Passage Booklet.

## 1a. Main Idea or Purpose Question 2

The primary purpose of the passage is to:

- A. explore Jackie's many facets.
- B. outline Jackie's early life.
- C. emphasize Jackie's uniqueness.
- D. investigate Jackie's popularity.

See next page for the *strategy-based explanation* of this question.

Chapter 5: Skill 1 – Foundations of Comprehension

---

### 1a. Main Idea or Purpose Question 2
### Strategy-based Explanation

The primary purpose of the passage is to:

A. explore Jackie's many facets.
B. outline Jackie's early life.
C. emphasize Jackie's uniqueness.
D. investigate Jackie's popularity.

---

This question asks you to discern the author's overall purpose in writing the passage. Although the question doesn't give you any information, the purpose of the passage would relate to its subject, which is clearly Jackie Bouvier Kennedy Onassis. Accordingly, a description of the passage's primary purpose would sum up many of the claims made about Jackie in the passage.

## Applying the Method

*Passage Excerpt*

**[P1]** Multifaceted and always elusive, **the former first lady never ceased to fascinate**; however, people had to be satisfied with only glimpses of this fashion icon, culture advocate, historic preservationist, polyglot, equestrienne, and book editor. Indeed, upon her death in 1994, **Jacqueline Kennedy Onassis was described as "the most intriguing woman in the world."** Often topping lists of the most admired individuals of the second half of the 20th century, this **celebrated woman** is likely someone many wish they had known. Barring such a possibility, the best way to fully appreciate **Jackie's exceptional nature** might be to consider the people she wished she had known.

**[P2] It is unsurprising that this woman who captured the public's imagination for decades distinguished herself from her peers early on.**

**[P3]** In a brief composition, Jackie provided deep insights into this bohemian threesome of poet, aesthete, and impresario with whom she strongly identified.

**[P4] The contest committee judged Jackie's essay to have exhibited a profound appreciation for the arts combined with a truly outstanding level of intellectual maturity and**

### Step 1: Ask Focus Questions

To help us determine the correct answer, we might begin by asking this version of a focus question: **What aspects of Jackie does the passage seem most concerned with?**

As we review the passage, we see Jackie portrayed in a consistent manner.

According to Paragraph 1, **Jackie never ceased to fascinate people.** She was said to be **the most intriguing woman in the world** as well as **a celebrated woman whose nature was exceptional**.

Paragraph 2 stresses that **even early on in her life Jackie stood out from her peers**.

Paragraph 3 reveals the contents of Jackie's Prix de Paris essay, while Paragraph 4 goes on to discuss how **the committee saw its contents**

**originality of thought.** Similarly, biographer Donald Spoto deemed Jackie "remarkably unorthodox," not unlike the men about whom she wrote in her unusual composition, which he pronounced "a masterpiece of perceptive improvisation." Thus, **from a pool of 1,279 applicants representing 224 colleges, Jacqueline Bouvier was declared the winner**.

[P5] Although Ms. Bouvier went on to decline the prestigious award, which would have involved living and working in Paris, she never gave up her dream of being the century's art director. As first lady, **she tirelessly promoted the arts and culture. Today, the John F. Kennedy Center for the Performing Arts in Washington, DC, is a legacy of Jackie's vision**.

as **evidence of her outstanding nature**. In fact, **Jackie would surpass a large number of applicants to win the scholarship contest.**

Paragraph 5 indicates Jackie continued to actively promote the arts, and **her efforts were publicly concretized in the building of the Kennedy Center for the Performing Arts**.

Given these direct claims, the passage seems most concerned with **showing that Jackie was exceptional and always stood out from others in some way**.

### Step 2: Compare Answer Choices with the Focus of the Passage

After considering our focus question, we determined that the passage seems most concerned with **showing that Jackie was exceptional and always stood out from others in some way.**

Looking at the answer choices, the phrase *emphasize Jackie's uniqueness* most closely corresponds to the focus of the passage on Jackie as exceptional and standing out. Therefore, **Choice C** is likely the correct answer.

### Step 3: Consider Your Chosen Answer as a Thesis or Conclusion

As a final test of our answer choice, we can ask whether **Choice C** could serve as a thesis statement or overall conclusion to the passage. Looking at Paragraph 1, the author states that Jackie never ceased to fascinate, was considered the most intriguing woman in the world when she died, and was generally a celebrated woman, all of which emphasize that she was exceptional or unique. Accordingly, a thesis statement explicitly *emphasiz[ing] Jackie's uniqueness* could easily fit into Paragraph 1. Therefore, we can feel confident that **Choice C** is the correct answer.

For an alternative method of explanation based more specifically on passage evidence, you can view this question in the UWorld Qbank (sold separately).

# Passage C: Probability and The Universe

The idea of probability is frequently misunderstood, in large part because of a conceptual confusion between objective probability and subjective probability. The failure to make this distinction leads to an erroneous conflation of genuine possibility with what is in fact merely personal ignorance of outcome. An example will clarify.

A standard die is rolled on a table, but the outcome of the roll is concealed. Should an observer be asked the chance that a particular number was rolled—five, say—the natural response is 1/6. However, this answer is incorrect. To say there is a one-in-six chance that a five was rolled implies there is an equal chance that any of the other numbers were rolled. But there is no equal chance, because the roll has already occurred. Hence, the probability that the result of the roll is a five is either 100% or 0%, and the same is true for each of the other numbers.

It might be objected that such an analysis is an issue of semantics rather than a substantive claim. For declaring the probability to be 1/6 is merely an expression that, for all we know, any number from 1 through 6 might have been rolled. But the difference between *for all we know* and what *is* remains crucially important, because it forestalls a tendency to make scientific assertions from a perspective biased by human perception....

[T]hus, it should be clear that claims about the purpose of the universe rest on shaky ground. In particular, we must be wary of inferences drawn from juxtaposing the existence of intelligent life with the genuinely improbable cosmological conditions that make such life possible. Joseph Zycinski provides an instructive account of those conditions: "If twenty billion years ago the rate of expansion were minimally smaller, the universe would have entered a stage of contraction at a time when temperatures of thousands of degrees were prevalent and life could not have appeared. If the distribution of matter were more uniform the galaxies could not have formed. If it were less uniform, instead of the presently observed stars, black holes would have formed from the collapsing matter." In short, the existence of life (let alone intelligent life) depends upon the initial conditions of the universe having conformed to an extremely narrow range of possible values.

It is tempting to go beyond Zycinski's factual point to draw deep cosmological, teleological, and even theological conclusions. But no such conclusions follow. The reason why a life-sustaining universe exists is that if it did not, there would be no one to wonder why a life-sustaining universe exists. This fact is a function not of purpose, but of pre-requisite. For instance, suppose again that a five was rolled on a die. We now observe a five on its face, not because the five was "meant to be" but because one side of the die had to land face up. Even if we imagine that the die has not six but six *billion* sides, that analysis is unaffected. Yes, it was unlikely that any given value would be rolled, but *some* value had to be rolled. That "initial condition" then set the parameters for what kind of events could possibly follow; in this case, the observation of the five.

Similarly, the existence of intelligent beings is evidence that certain physical laws obtained in the universe. However, it is not evidence that those beings or laws were necessary or intended rather than essentially random. The subjective probability of that outcome is irrelevant, because in this universe the objective probability of its occurrence is 100%. Thus, the seemingly low initial chance that such a universe would exist is not in itself indicative of a purpose to its existence.

*Probability and The Universe* ©UWorld

Annotations for this passage can be found in the CARS Passage Booklet.

## 1a. Main Idea or Purpose Question 3

The author's main point regarding the existence of the universe is that:

A. facts about the universe do not justify inferences about its purpose.
B. the initial conditions of the universe should guide how we understand other facts about it.
C. facts about the universe are inevitably framed by biases of human perception.
D. facts about the universe depend on a subjective understanding of probability.

See next page for the *strategy-based explanation* of this question.

Chapter 5: Skill 1 – Foundations of Comprehension

### 1a. Main Idea or Purpose Question 3
### Strategy-based Explanation

The author's main point regarding the existence of the universe is that:

A. facts about the universe do not justify inferences about its purpose.
B. the initial conditions of the universe should guide how we understand other facts about it.
C. facts about the universe are inevitably framed by biases of human perception.
D. facts about the universe depend on a subjective understanding of probability.

The question asks us to consider what the author says specifically about the existence of the universe. This information is useful because the first three paragraphs of the passage focus on a different topic, objective vs. subjective probability. Therefore, when approaching this question, we can narrow down which parts of the passage are most relevant to review.

## Applying the Method

*Passage Excerpt*

**[P4]** [T]hus, it should be clear that **claims about the purpose of the universe rest on shaky ground**. In particular, we must be wary of inferences drawn from juxtaposing the existence of intelligent life with the genuinely improbable cosmological conditions that make such life possible.

**[P5]** The reason why a life-sustaining universe exists is that if it did not, there would be no one to wonder why a life-sustaining universe exists. **This fact is a function not of purpose, but of pre-requisite.** For instance, suppose again that a five was rolled on a die. We now observe a five on its face, **not because the five was "meant to be"** but because one side of the die had to land face up. Even if we imagine that the die has not six but six *billion* sides, that analysis is unaffected. Yes, it was unlikely that any given value would be rolled, but *some* value had to be rolled.

**[P6]** Similarly, the existence of intelligent beings is evidence that certain physical laws obtained in the universe. However, **it is not evidence that those beings or laws were necessary or intended rather than essentially random**. The subjective probability of that outcome is irrelevant, because in this universe

**Step 1: Ask Focus Questions**

Applying one of our focus questions, we might begin by asking: **What does the author primarily want people to know about the existence of the universe?**

Paragraphs 4, 5, and 6 all discuss the existence of the universe. Reviewing these paragraphs, the discussion can be broken into two themes: the probability that the universe would have existed and **whether the universe has a purpose**.

Probability:

These paragraphs all convey the idea that **the universe's existence was an unlikely occurrence**. Paragraph 4 calls the existence of the universe genuinely improbable; Paragraph 5 compares it to a one-in-six-billion die roll; and Paragraph 6 refers to the seemingly low initial chance that the universe would have existed.

Purpose:

The paragraphs also stress that **we can't conclude that the universe has a purpose**. For instance, Paragraph 4 states that **claims about the purpose of the universe rest on shaky ground**. Similarly, Paragraph 5 warns against thinking of outcomes as **"meant to be,"** and Paragraph 6 suggests that the universe might have been **essentially random**.

the objective probability of its occurrence is 100%. Thus, **the seemingly low initial chance that such a universe would exist is not in itself indicative of a purpose to its existence**.

> Given these consistent claims, what the author primarily wants people to know about the existence of the universe seems to be:
>
> **The existence of the universe was improbable, but that doesn't mean there is a purpose to its existence.**

### Step 2: Compare Answer Choices with the Focus of the Passage

After considering our focus question, we have concluded the author most wants people to know that **the existence of the universe was improbable, but that doesn't mean there is a purpose to its existence**. As we review the answer choices, we see that **Choice A** seems to represent a more general version of that statement: *facts about the universe do not justify inferences about its purpose*. In addition, this is the only answer choice that refers to the *purpose* of the universe, which we noted was an important theme for the author. Therefore, **Choice A** is most likely correct.

### Step 3: Consider Your Chosen Answer as a Thesis or Conclusion

To double-check this answer, we can apply our final step: could **Choice A** serve as a thesis statement or overall conclusion to the passage? In fact, we can see that the final sentence of the passage makes a similar point about the purpose of the universe: **the seemingly low initial chance that such a universe would exist is not in itself indicative of a purpose to its existence**. Accordingly, the statement *facts about the universe do not justify inferences about its purpose* could easily be a conclusion to the passage. Thus, we can confirm that **Choice A** is the correct answer.

For an alternative method of explanation based more specifically on passage evidence, you can view this question in the UWorld Qbank (sold separately).

Lesson 5.2
# CMP Subskill 1b. Meaning of Term

| Skill 1: Foundations of Comprehension ||
|---|---|
| Subskill | Student Objective |
| 1a. Main Idea or Purpose | Identify the central theme or purpose of the passage |
| **1b. Meaning of Term** | **Determine the meaning of a term as used in the passage** |
| 1c. Direct Passage Claims | Identify claims made directly in the passage |
| 1d. Implicit Claims or Assumptions | Identify implicit claims or unstated assumptions in the passage |
| 1e. Identifying Passage Perspectives | Identify perspectives or positions presented in the passage |
| 1f. Further Implications of Passage Claims | Recognize further implications of claims stated in the passage |

**1b. Meaning of Term** questions require you to figure out what the author (or a quoted source) means by a specific word or phrase. In particular, you must determine how the term is *used in the context of the passage*. Thus, answering these questions is not just a matter of giving a correct definition. Rather, you must analyze how the term contributes to the point being made where it appears in the passage.

Nevertheless, since these questions do ask what a term means, you might wonder whether they are basically vocabulary questions. If so, wouldn't that mean that answering them unfairly depends on outside knowledge? However, remember that a term may be an entire phrase, not just an individual word. More importantly, knowing what a term typically means does not necessarily tell you how it is being used in a particular passage, or what the author hopes to convey by using it. Therefore, although external knowledge of vocabulary can be helpful, answering these types of questions must still be based on passage information.

### Why Can't They Just Tell Me Where It Is?

While some **1b. Meaning of Term** questions include a paragraph reference, others simply mention the term without indicating where it appears in the passage. Unfortunately, the "Ctrl+F" command does not work on the MCAT. However, keep in mind that when a question asks about a term it probably wasn't chosen at random; it likely stands out in some way. Accordingly, these questions are often drawn from:

- **unusual words or jargon**
    "mobocracy"; "linguistic currency"; "dilemma of symbolic identity"
- **quoted material**
    "Johnston's claim that the Lacanian ego is 'a coagulation of inter- and trans-subjective alien influences'"
- **metaphorical or poetic-sounding language**
    "found consolation in the thought that his fate had not been his fate"
- **points that the author specifically emphasizes**
    "there is a sense in which—if not objectively, then at least concretely—representational art is superior to abstract art."

Such terms often affect how the passage should be understood, and it is worth taking mental note when you encounter them. As you continue to practice reading CARS passages, you will likely find this process happening more naturally, making it easier to recall where terms appear in the text.

Once you have located the term, the following method will enable you to determine its meaning as used in the passage.

## Method for Answering 1b. Meaning of Term Questions

### Step 1: Review the Context in Which the Term Is Used

The first step in determining the meaning of a word or phrase is to review the context in which it appears. Generally, it is helpful to reread not only the sentence containing the term, but the ones preceding and following it. This process will refamiliarize you with what is being discussed where the term is used.

### Step 2: Consider the Role of the Term in the Passage

Second, consider what the passage is doing when the term appears. For instance, is it clarifying a previous statement or making a new point? Is it expressing an attitude or explaining some phenomenon? By considering what an author is trying to accomplish where a term is used, you can discern how the term functions to serve that goal.

**Step 3: Ask What the Term Adds**

Finally, a useful trick for figuring out the meaning of a term is to ask: What does this term add to the sentence in which it appears?  Or, approaching that question from the other direction: What would be missing from the sentence if this term were left out?  This process can help you to isolate the term's effect on the sentence.

> **Subskill 1b. Meaning of Term**
> **(3 Practice Questions)**
>
> We now look at three examples from the UWorld Qbank. In each case, the passage and question are first given without commentary, allowing you to practice applying the method yourself. Then, the passage and question are presented again, this time with annotations of the passage (see the CARS Passage Booklet) and a step-by-step explanation of the question using the described method.

# Passage D: American Local Motives

Locomotives were invented in England, with the first major railroad connecting Liverpool and Manchester in 1830. However, it was in America that railroads would be put to the greatest use in the nineteenth century. On May 10, 1869, the Union Pacific and Central Pacific lines met at Promontory Point, Utah, joining from opposite directions to complete a years-long project—the Transcontinental Railroad. This momentous event connected the eastern half of the United States with its western frontier and facilitated the construction of additional lines in between. As a result, journeys that had previously taken several months by horse and carriage now required less than a week's travel. By 1887 there were nearly 164,000 miles of railroad tracks in America, and by 1916 that number had swelled to over 254,000.

While the United States still has the largest railroad network in the world, it operates largely in the background of American life, and citizens no longer view trains with the sense of importance those machines once commanded. Nevertheless, the economic and industrial advantages those citizens enjoy today would not have been possible without America's history of trains; as Tom Zoellner reminds us, "Under the skin of modernity lies a skeleton of railroad tracks." Although airplanes and automobiles have now assumed greater prominence, the time has arrived for the resurgence of railroads. A revitalized and advanced railway system would confer numerous essential benefits on both the United States and the globe.

The chief obstacles to garnering support for such a project are the current dominance of the automobile and the languishing technology of existing railroads. In a sense these two obstacles are one, as American dependence on personal automobiles is partially due to the paucity of rapid public transportation. The railroads of Europe and Japan, by comparison, have vastly outpaced their American counterparts. Japan has operated high-speed rail lines continuously since 1964, and in 2007, a French train set a record of 357 miles per hour. While that speed was achieved under tightly controlled conditions, it still speaks to the great disparity in railroad development between the United States and other countries since the mid-twentieth century. British trains travel at speeds much higher than those in America, where both the trains themselves and the infrastructure to support them have simply been allowed to fall behind. In much of Europe it is common for trains to travel at close to 200 miles per hour.

To invest in a modern network of railroads would improve the United States in much the same way that the first railroads did in the nineteenth and early twentieth centuries. A high-speed passenger rail system would dramatically transform American life as travel between cities and states became quicker and more convenient, encouraging commerce, business, and tourism. Such a system would also make important strides in environmental preservation. According to a 2007 British study, "CO2 emissions from aircraft operations are...at least five times greater" than those from high-speed trains. For similar reasons, Osaka, Japan, was ranked as "the best…green transportation city in Asia" by the 2011 *Green City Index*. As Lee-in Chen Chiu notes in *The Kyoto Economic Review*, Osakans travel by railway more than twice as much as they travel by car.

It is true that developing a countrywide high-speed rail system would come with significant costs. However, that was also true of the original Transcontinental Railroad, as indeed it is with virtually any great project undertaken for the public good. We should thus move ahead with confidence that the rewards will outweigh the expenditure as citizens increasingly choose to travel by train. Both for society's gain and the crucial well-being of the planet, our path forward should proceed upon rails.

*American Local Motives* ©UWorld

Annotations for this passage can be found in the CARS Passage Booklet.

### 1b. Meaning of Term Question 1

In the context of the passage, which of the following best explains Zoellner's claim that "[u]nder the skin of modernity lies a skeleton of railroad tracks"?

A. American trains now represent an obsolete technology.
B. American trains have been overshadowed by other forms of transportation.
C. American trains have been largely retired from use in current society.
D. American trains have been crucial in producing the country as we know it.

See next page for the *strategy-based explanation* of this question.

> ## 1b. Meaning of Term Question 1
> ## Strategy-based Explanation
>
> In the context of the passage, which of the following best explains Zoellner's claim that "[u]nder the skin of modernity lies a skeleton of railroad tracks"?
>
> A. American trains now represent an obsolete technology.
> B. American trains have been overshadowed by other forms of transportation.
> C. American trains have been largely retired from use in current society.
> D. American trains have been crucial in producing the country as we know it.

The question does not include a paragraph reference, but the term is a quotation. So, even if you don't remember exactly where Zoellner's claim appears in the passage, it can be located fairly quickly by looking for quoted text. In this case, you can find it in Paragraph 2.

## Applying the Method

*Passage Excerpt*

[P2] While the United States still has the largest railroad network in the world, it operates largely in the background of American life, and citizens no longer view trains with the sense of importance those machines once commanded. Nevertheless, the economic and industrial advantages those citizens enjoy today would not have been possible without America's history of trains; **as Tom Zoellner reminds us, "Under the skin of modernity lies a skeleton of railroad tracks."** Although airplanes and automobiles have now assumed greater prominence, the time has arrived for the resurgence of railroads.

### Step 1: Review the Context in Which the Term Is Used

Looking at the surrounding sentences, we can see that the context for the term is a contrast between the current and past status of railroads in America. The author first remarks that railroads are not as prominent in American society as they once were, then states that they should become prominent again. Thus, **Zoellner's claim** helps to connect these ideas.

### Step 2: Consider the Role of the Term in the Passage

More specifically, **Zoellner's claim** appears as the second half of a longer sentence. Thus, by stating "**as Tom Zoellner reminds us**," the author indicates that they are using Zoellner's quotation to emphasize the preceding part of that sentence. Therefore, the role of Zoellner's claim is to further convey the idea that "the economic and industrial advantages those citizens enjoy today would not have been possible without America's history of trains."

### Step 3: Ask What the Term Adds

We can further test our conclusion about the role of the term by mentally removing **Zoellner's claim** from the sentence in which it appears. That sentence would then become:

"Nevertheless, the economic and industrial advantages those citizens enjoy today would not have been possible without America's history of trains; as Tom Zoellner reminds us."

This is another way to see that **Zoellner's claim** is intended to stress the point the author has just made. In other words, what the term adds is a reinforcement of the author's assertion that trains were essential to creating modern American society.

Turning to the answer choices, we see that **Choice D** is a paraphrase of this same idea: *American trains have been crucial in producing the country as we know it*. Therefore, **Choice D** should be the correct answer.

For an alternative method of explanation based more specifically on passage evidence, you can view this question in the UWorld Qbank (sold separately).

---

**Note: Content vs. Structure**

You may have noticed in using this method, especially with Step 3, that we were able to answer the question without actually having to interpret the quote itself. Based solely on the structure of the sentence, we could tell what Zoellner's claim was supposed to convey—regardless of whether we understood the content of that claim. You will likely find something similar is true of many CARS questions. By approaching them in a logical way, you can answer them correctly even in cases where a passage seems difficult to understand.

# Passage E: The Divine Sign of Socrates

From his bare feet to his bald pate, the potentially shapeshifting figure of Socrates found in the literary tradition that arose after his controversial trial and death presents an intriguing array of oddities and unorthodoxies. Most conspicuously, his unshod and shabby sartorial state flaunted poverty at a time when the city of Athens had become obsessed with wealth and its trappings. Yet the philosopher's peculiar appearance was but a hint of the strange new calling he embraced. Inspired perhaps by the famous Delphic dictum "Know thyself," he embarked on a mission devoted to finding truth through dialogue. In what struck some as a dangerous new method of inquiry, he subjected nearly everyone he encountered to intense cross-examination, mercilessly exposing the ignorance of his interlocutors. Moreover, in a culture that still put stock in magic, the highly charismatic, entertaining, and at times infuriating Socrates appeared to be a sorcerer bewitching the aristocratic young men of Athens who followed him fanatically about the agora.

By all credible accounts, this exceedingly eccentric, self-styled radical truth-seeker had more than a whiff of the uncanny about him. As Socrates himself explains in Plato's *Republic*, he was both blessed and burdened with a supernatural phenomenon in the form of a *daimonion* or inner spirit that always guided him: "This began when I was a child. It is a voice, and whenever it speaks, it turns me away from something I am about to do, but it never encourages me to do anything." An overtly rational thinker, Socrates nonetheless considered these warnings—or, in James Miller's words, "the audible interdictions he experienced as irresistible"—to be infallible. Such oracular injunctions were highly anomalous as tutelary spirits were thought to assume a more nuanced presence. Some scholars have dismissed Socrates' recurring sign as a hallucination or psychological aberration. Others have conjectured that the internal voice might be attributable to the cataleptic or trancelike episodes from which the philosopher purportedly suffered. Indeed, as Miller notes, "Socrates was storied for the abstracted states that overtook him"; not infrequently, his companions would see him stop in his tracks and stand still for hours, completely lost in thought.

As Socrates further insisted, it was only the protestations of this apotreptic voice that held him back from entering the political arena. Even so, its personal admonitions could not spare him persecution. Despite the political amnesty extended by the resurgent democracy that succeeded the interim pro-Spartan oligarchy, the thinker's notoriety and ambiguous allegiances aroused suspicions. In 399 BCE, Socrates was brought before the court on trumped-up charges of impiety; these included willfully neglecting the traditional divinities, flagrantly introducing new gods to the city, and wittingly corrupting the youth. Athenian society recognized no division between religious and civic duties, and capricious gods demanded constant appeasement through sacrifices and rituals. Consequently, belief in a purely private deity—particularly a wholly benevolent deity conveying unequivocal messages—was inadmissible. Worse, as Socrates' own testimony revealed, he honored this personal god's authority above even the laws of the city. Hence, the philosopher's *daimonion* loomed over his indictment, conviction, and sentencing.

Nevertheless, in his defense speech as reconstructed by Plato in the *Apology*, Socrates maintained confidence in the protective nature and prophetic powers of his inner monitor. He never questioned its affirmatory silence toward his predicament, remarking, "The divine faculty would surely have opposed me had I been going to evil and not to good." Thus, Socrates acknowledged that his *daimonion* had its reasons, however inscrutable. Variously described as malcontent and martyr, public nuisance and prophet, laughingstock and hero, the mercurial Athenian, like the sign that guided him, was difficult to fathom yet impossible to ignore.

*The Divine Sign of Socrates* ©UWorld

Annotations for this passage can be found in the CARS Passage Booklet.

### 1b. Meaning of Term Question 2

As the passage author uses the term, *apotreptic* most likely means:

A. strongly discouraging.
B. irritatingly inconvenient.
C. highly ambiguous.
D. seemingly arbitrary.

See next page for the *strategy-based explanation* of this question.

# Chapter 5: Skill 1 – Foundations of Comprehension

---

### 1b. Meaning of Term Question 2
### Strategy-based Explanation

As the passage author uses the term, *apotreptic* most likely means:

A. strongly discouraging.
B. irritatingly inconvenient.
C. highly ambiguous.
D. seemingly arbitrary.

---

This question does not include a paragraph reference. However, since "apotreptic" is an unusual word, you likely noticed it as you were reading. The term may also have stood out because it appears in the first sentence in Paragraph 3, which offers an example of Socrates' inner voice to support the description in Paragraph 2.

## Applying the Method

*Passage Excerpt*

**Step 1: Review the Context in Which the Term Is Used**

[P2] As Socrates himself explains in Plato's *Republic*, he was both blessed and burdened with a supernatural phenomenon in the form of a *daimonion* or inner spirit that always guided him: "This began when I was a child. It is a voice, and whenever it speaks, it turns me away from something I am about to do, but it never encourages me to do anything." An overtly rational thinker, Socrates nonetheless considered these warnings—or, in James Miller's words, "the audible interdictions he experienced as irresistible"—to be infallible. Such oracular injunctions were highly anomalous as tutelary spirits were thought to assume a more nuanced presence. Some scholars have dismissed Socrates' recurring sign as a hallucination or psychological aberration. Others have conjectured that the internal voice might be attributable to the cataleptic or trancelike episodes from which the philosopher purportedly suffered. Indeed, as Miller notes, "Socrates was storied for the abstracted states that overtook him"; not infrequently, his companions would see him stop in his tracks and stand still for hours, completely lost in thought.

The context in which the term "**apotreptic**" appears comprises a description of Socrates' inner voice and its effect on his life. From Paragraph 2 we learn that this voice functioned by warning Socrates not to do certain things.

Paragraph 3 then provides a specific example of how this inner voice stopped Socrates from entering politics.

**[P3]** As Socrates further insisted, it was only the protestations of **this apotreptic voice** that held him back from entering the political arena. Even so, its personal admonitions could not spare him persecution. Despite the political amnesty extended by the resurgent democracy that succeeded the interim pro-Spartan oligarchy, the thinker's notoriety and ambiguous allegiances aroused suspicions.

**Step 2: Consider the Role of the Term in the Passage**

The sentence in which "**apotreptic**" appears forms a transition from the description of Socrates' inner voice in Paragraph 2 to an example of the effect it had on him in Paragraph 3. Within this sentence, the term "**apotreptic**" is attached to the word "**voice**." Thus, the role of the term is to describe Socrates' guiding voice and to connect the information in Paragraph 2 to the example in Paragraph 3.

### Step 3: Ask What the Term Adds

To further test how the term "**apotreptic**" is used, let's see what happens if we remove it from the phrase it appears in. We are then left with the statement:

"As Socrates further insisted, it was only the protestations of this voice that held him back from entering the political arena."

We can now see that nothing substantial is lost by removing "**apotreptic**": the sentence still makes perfect sense and performs the same role. Accordingly, the term does not seem to add any new information about Socrates' inner voice; instead, "**apotreptic**" most likely just refers to the previous description of that voice.

Thus, what the term adds is simply a reference to the existing information about Socrates' inner voice given in Paragraph 2. As we have noted, this voice is described as always forbidding Socrates from taking certain actions, for example, entering politics. Of the answer choices, only **Choice A**, *strongly discouraging*, reflects this description. Therefore, **Choice A** should be correct.

For an alternative method of explanation based more specifically on passage evidence, you can view this question in the UWorld Qbank (sold separately).

## Passage B: Jackie in 500 Words

Born in New York in 1929, Jacqueline Bouvier first came into the public eye as the wife of the 35th president of the United States, John F. Kennedy. The president was assassinated in 1963, and by the end of what turned out to be a turbulent decade, Mrs. Kennedy had transformed herself into the enigmatic Jackie O., wife of Greek shipping magnate Aristotle Onassis. Multifaceted and always elusive, the former first lady never ceased to fascinate; however, people had to be satisfied with only glimpses of this fashion icon, culture advocate, historic preservationist, polyglot, equestrienne, and book editor. Indeed, upon her death in 1994, Jacqueline Kennedy Onassis was described as "the most intriguing woman in the world." Often topping lists of the most admired individuals of the second half of the 20th century, this celebrated woman is likely someone many wish they had known. Barring such a possibility, the best way to fully appreciate Jackie's exceptional nature might be to consider the people she wished she had known.

It is unsurprising that this woman who captured the public's imagination for decades distinguished herself from her peers early on. Notably, in 1951, Ms. Bouvier entered a scholarship contest sponsored by *Vogue* and open to young women in their final undergraduate year, the annual Prix de Paris. Among other assignments, applicants were asked to compose a 500-word essay, "People I Wish I Had Known," spotlighting three individuals influential in art, literature, or culture. The future first lady chose an iconoclastic trio from the Victorian era: the French symbolist poet Charles Baudelaire, the Irish wit Oscar Wilde, and the innovative Ballets Russes dance company founder Sergei Diaghilev.

In a brief composition, Jackie provided deep insights into this bohemian threesome of poet, aesthete, and impresario with whom she strongly identified. She concluded that Baudelaire deployed "venom and despair" as "weapons" in his poetry. She idolized Wilde for being able "with the flash of an epigram to bring about what serious reformers had for years been trying to accomplish." Diaghilev she defined as an artist of a different sort, someone who "possessed what is rarer than artistic genius in any one field—the sensitivity to take the best of each man and incorporate it into a masterpiece." As Jackie poignantly observed, such a work is "all the more precious because it lives only in the minds of those who have seen it," dissipating soon after. Furthermore, although these men espoused different disciplines, she discerned that "a common theory runs through their work, a certain concept of the interrelation of the arts." Finally, foreshadowing her self-assumed role in the White House as the nation's unofficial minister of the arts, Jackie paid homage with her vision: "If I could be a sort of Overall Art Director of the Twentieth Century, watching everything from a chair hanging in space, it is their theories of art that I would apply to my period."

The contest committee judged Jackie's essay to have exhibited a profound appreciation for the arts combined with a truly outstanding level of intellectual maturity and originality of thought. Similarly, biographer Donald Spoto deemed Jackie "remarkably unorthodox," not unlike the men about whom she wrote in her unusual composition, which he pronounced "a masterpiece of perceptive improvisation." Thus, from a pool of 1,279 applicants representing 224 colleges, Jacqueline Bouvier was declared the winner.

Although Ms. Bouvier went on to decline the prestigious award, which would have involved living and working in Paris, she never gave up her dream of being the century's art director. As first lady, she tirelessly promoted the arts and culture. Today, the John F. Kennedy Center for the Performing Arts in Washington, DC, is a legacy of Jackie's vision.

*Jackie in 500 Words* ©UWorld

Annotations for this passage can be found in the CARS Passage Booklet.

## 1b. Meaning of Term Question 3

In the context of the passage, Spoto's phrase "remarkably unorthodox" (Paragraph 4) suggests that Jackie's scholarship application:

- A. displayed her surprising viewpoints.
- B. reflected her unconventional education.
- C. bent the contest rules.
- D. portrayed individuals who seemed unlike her.

See next page for the *strategy-based explanation* of this question.

Chapter 5: Skill 1 – Foundations of Comprehension

> **1b. Meaning of Term Question 3**
> **Strategy-based Explanation**
>
> In the context of the passage, Spoto's phrase "remarkably unorthodox" (Paragraph 4) suggests that Jackie's scholarship application:
>
> A. displayed her surprising viewpoints.
> B. reflected her unconventional education.
> C. bent the contest rules.
> D. portrayed individuals who seemed unlike her.

This question is phrased a little differently than the other **1b. Meaning of Term** questions we have seen, but it is still asking how the term is used in the passage. What Spoto's phrase suggests about Jackie's application is a matter of what idea in the passage his quote is being used to convey.

## Applying the Method

*Passage Excerpt*

**[P4]** The contest committee judged Jackie's essay to have exhibited a profound appreciation for the arts combined with a truly outstanding level of intellectual maturity and originality of thought. **Similarly**, biographer Donald Spoto deemed Jackie **"remarkably unorthodox,"** not unlike the men about whom she wrote in her unusual composition, which he pronounced "a masterpiece of perceptive improvisation." Thus, from a pool of 1,279 applicants representing 224 colleges, Jacqueline Bouvier was declared the winner.

### Step 1: Review the Context in Which the Term Is Used

Looking at the sentences surrounding the term "**remarkably unorthodox**" in Paragraph 4, we can see that the context is the evaluation of Jackie's scholarship essay and the outstanding qualities it displayed. The author cites Spoto's words in the course of describing how Jackie's essay won the contest based on its artistic sensibility, maturity, and originality.

### Step 2: Consider the Role of the Term in the Passage

Spoto's quote is introduced with the word "**Similarly**." Accordingly, the author is indicating that the phrase "**remarkably unorthodox**" reflects some or all of the same qualities that the contest committee used to describe Jackie's essay and that Spoto now attributes to Jackie herself.

Thus, the role of the term "**remarkably unorthodox**" is to encapsulate and reinforce the contest committee's preceding description of Jackie's essay.

## Step 3: Ask What the Term Adds

We can confirm our conclusion about "**remarkably unorthodox**" in the previous step by removing it from the phrase in which it appears. We are now left with the partial statement:

"Similarly, biographer Donald Spoto deemed Jackie…"

Based on this phrase, we know that whatever follows "Jackie" must convey something similar to what was said before. Hence, what the term adds is the way in which Spoto's quote was like the contest committee's evaluation. Accordingly, Spoto's phrase "**remarkably unorthodox**" suggests that Jackie possessed one or more of the following qualities: appreciation for art; intellectual maturity; or originality of thought.

Looking at the answer choices, we can see that only **Choice A** reflects any of these qualities. Specifically, if Jackie's essay showed originality of thought, then her views must have been unexpected—in other words, her essay would have *displayed her surprising viewpoints*. Accordingly, **Choice A** matches what Spoto's phrase suggests and so is the correct answer.

For an alternative method of explanation based more specifically on passage evidence, you can view this question in the UWorld Qbank (sold separately).

Lesson 5.3
# CMP Subskill 1c. Direct Passage Claims

| Skill 1: Foundations of Comprehension ||
|---|---|
| Subskill | Student Objective |
| 1a. Main Idea or Purpose | Identify the central theme or purpose of the passage |
| 1b. Meaning of Term | Determine the meaning of a term as used in the passage |
| **1c. Direct Passage Claims** | **Identify claims made directly in the passage** |
| 1d. Implicit Claims or Assumptions | Identify implicit claims or unstated assumptions in the passage |
| 1e. Identifying Passage Perspectives | Identify perspectives or positions presented in the passage |
| 1f. Further Implications of Passage Claims | Recognize further implications of claims stated in the passage |

To answer **1c. Direct Passage Claims** questions, you must identify and understand what the passage explicitly says. As every question involves understanding what the passage says, why is such an obvious point being highlighted here? The reason is that in these cases, nothing else is required. In other words, these are questions whose answers are based solely on recognizing passage claims, not engaging in any further reasoning.

## Method for Answering 1c. Direct Passage Claims Questions

### Step 1: Verify What the Passage Directly States

Ideally, if you have read the passage carefully, then you understand its major themes. However, it can be important to recheck the relevant parts of the passage to make sure you aren't misremembering something or forgetting a particular detail. In addition, since these questions concern specific statements in the passage, this process gives us an idea of what the correct answer should look like.

### Step 2: Compare Answer Choices with Those Statements

The next step is to find the answer choice that matches what you have verified about passage statements. In many cases, the correct answer will use exact or nearly exact wording from the passage. If not, the correct answer choice will state the same thing in different words. Either way, only one answer will accurately reflect what the passage says about the question topic.

Given how straightforward this method is, you might be thinking that these types of questions seem simple. And oftentimes, they are—it would be fair to call them the most basic type of CARS question. (They are also one of the most common.) Nevertheless, challenges can still arise in answering them. For instance, you might be tripped up by:

- claims that do appear in the passage but are irrelevant to the question being asked;
- claims that would be consistent with the author's view, but that the author does not actually express;
- questions using "NOT" or "EXCEPT," which may require more extensive review of the passage.

These obstacles can cause problems in multiple types of questions, not just **1c. Direct Passage Claims**. They may stand out more with these questions, however, because they disrupt what should otherwise be an uncomplicated process of identifying statements made in the passage.

### But How Can I Be Sure It's a 1c Question?

A further challenge is that sometimes the exact question type is not obvious from the way it is worded. For instance, the question "Which of the following statements is consistent with the passage?" might be based on explicit passage information, but it could also refer to something implied. So, how can you tell from the start that the method for **1c. Direct Passage Claims** questions is the right one?

In certain cases, you may not be able to tell immediately. However, even if it turns out that the question is asking for an assumption or inference instead of a direct claim, you will still need to verify what the passage directly says. Thus, you can think of this method as a starting point that the methods for some other subskills build on. This relationship is illustrated further in Lessons 5.4 and 5.6, where the methods for addressing **1d. Implicit Claims or Assumptions** and **1f. Further Implications of Passage Claims** are similar to this one but include additional steps.

## Subskill 1c. Direct Passage Claims
## (5 Practice Questions)

We now look at five examples from the UWorld Qbank. In each case, the passage and question are first given without commentary, allowing you to practice applying the method yourself. Then, the passage and question are presented again, this time with annotations of the passage (see the CARS Passage Booklet) and a step-by-step explanation of the question using the described method.

## Passage F: The Victorian Internet

Lasting roughly from 1820 to 1914, the Victorian Era is often defined by its many distinctive sociological conditions, including industrialization, urbanization, railroad travel, imperialism, territorial expansionism, and the frictions sparked by Darwinism and democratic reform. However, descriptors that may not come as readily to mind are "the Information Age" and "the Age of Communication"; nor would we likely associate this era with something called the "highway of thought," which emanated from the electric telegraph—an invention that, in retrospect, could be renamed "the Victorian Internet." Yet the rise of the electric telegraph in the mid-1800s constituted the greatest revolution in communication since the invention of the printing press in the fifteenth century and until the launch of the World Wide Web at the end of the twentieth century.

Historically, messages could be conveyed only as fast as a person could travel from one location to another. However, by the end of the eighteenth century, the Chappe brothers had constructed a rudimentary optical telegraph on a hilltop tower in France. Using the large, jointed arms attached to the roof, operators could form various configurations to communicate a message to a similar tower farther away, which would then relay the message to a third tower, and so on, in a long-distance chain. Nevertheless, even with telescopes, visibility severely hampered the efficacy of this semaphore system. After countless attempts by scientists to use electricity to transmit messages, innovators in both Britain and America, including the painter and polymath Samuel F.B. Morse, worked at harnessing electromagnetic forces to send communications via cable. But while in Britain the technology was initially reserved for railway signaling, in the U.S. it culminated in the transmission of a message along a 40-mile telegraphic wire from Washington, D.C., to Baltimore in 1844. Using a binary code of short and long electrical impulses, or "dots and dashes," Morse dispatched the words: "What hath God wrought?"

Within 20 years, telegraph cables crisscrossed the continental U.S. and much of Europe. The telegraph office became ubiquitous, and telegrams—often bouncing from one office to another like emails tossed from server to server—reached ever more remote destinations. Following some spectacular failures, a durable cable was stretched across the floor of the Atlantic Ocean, enabling Europe and the U.S. to exchange messages within minutes. In the 1870s, the British Empire connected London with outposts in India and Australia. While individual telegram speed remained relatively constant, an endless deluge of information began to pour through the wires. There was, as Tom Standage observes, "an irreversible acceleration in the pace of business life," reflecting how "telegraphy and commerce thrived in a virtuous circle." It was as though "rapid long-distance communication had effectively obliterated time and space," begetting the phenomenon known as "globalization."

A new type of skilled worker, the telegrapher, was born. He or she belonged to a vast, online community, whose semi-anonymous members shared a unique intermediary role as well as a language of dots and dashes—vaguely prefiguring the exchange of bits and bytes along modern computer networks. Like today's online communities, these telegraphers fostered their own subculture: jokes and anecdotes flew over the wires; some operators invented systems to play games; and occasional romances blossomed. Regrettably, hackers, scammers, and shady entrepreneurs also frequented the byways of this early internet.

Predictably, newer forms of the original telegraphic technology—first the telephone and, much later, the fax machine—eventually encroached. The last telegram departed from India in 2013, but the twenty-first century has arguably seen a revival of this communication form in the text message. As Standage claims, texting has not only "reincarnate[d] a defunct nineteenth-century technology" but reinforced "the democratization of telecommunications" inaugurated by the miraculous Victorian Internet.

*The Victorian Internet* ©UWorld

Annotations for this passage can be found in the CARS Passage Booklet.

### 1c. Direct Passage Claims Question 1

Given passage information, which of the following assertions most aptly describes how the economic climate of the Victorian Era was transformed as a result of the telegraph?

A. It adapted to regulations controlling the pace of business.
B. It adapted to an acceleration of foreign communications.
C. It adapted to an increase in the localization of markets.
D. It adapted to a reduction in domestic trade deals.

See next page for the *strategy-based explanation* of this question.

## 1c. Direct Passage Claims Question 1
## Strategy-based Explanation

Given passage information, which of the following assertions most aptly describes how the economic climate of the Victorian Era was transformed as a result of the telegraph?

A. It adapted to regulations controlling the pace of business.
B. It adapted to an acceleration of foreign communications.
C. It adapted to an increase in the localization of markets.
D. It adapted to a reduction in domestic trade deals.

This complicated-looking question could be rephrased as "How did Victorian economics change because of the telegraph?" We then simply need to find where the passage discusses the Victorian economy and verify what it says about the telegraph's impact.

## Applying the Method

*Passage Excerpt*

**Step 1: Verify What the Passage Directly States**

[P3] Following some spectacular failures, a durable cable was stretched across the floor of the Atlantic Ocean, enabling Europe and the U.S. to exchange messages within minutes. In the 1870s, the British Empire connected London with outposts in India and Australia. While individual telegram speed remained relatively constant, an endless deluge of information began to pour through the wires. There was, as Tom Standage observes, "**an irreversible acceleration in the pace of business life,**" reflecting how "**telegraphy and commerce thrived in a virtuous circle.**" It was as though "rapid long-distance communication had effectively obliterated time and space," begetting the phenomenon known as "**globalization.**"

The Victorian economy is discussed in Paragraph 3. The author cites Tom Standage's account of how the telegraph caused **an irreversible acceleration** in business activity. Telegraphy and commerce mutually reinforced each other as they **thrived in a virtuous circle**, leading to an early form of **globalization**.

### Step 2: Compare Answer Choices with Those Statements

As we have seen, the passage mentions three effects of the telegraph on the Victorian economy: it caused an overall **acceleration in the pace of conducting business,** it led to a **mutual thriving of business and telegraphy**, and it helped to bring about **globalization**.

Similarly, **Choice B**, *an acceleration of foreign communications*, refers to both **an acceleration of pace** and an aspect of **globalization**. Therefore, **Choice B** reflects the relevant passage claims and is the correct answer.

For an alternative method of explanation based more specifically on passage evidence, you can view this question in the UWorld Qbank (sold separately).

## Passage G: Food Costs and Disease

Because frequent consumption of unhealthy foods is strongly linked with cardiometabolic diseases, one way for governments to combat those afflictions may be to modify the eating habits of the general public. Applying economic incentives or disincentives to various types of foods could potentially alter people's diets, leading to more positive health outcomes.

Utilizing national data from 2012 regarding food consumption, health, and economic status, Peñalvo et al. concluded that such price adjustments would help to prevent deaths related to cardiometabolic diseases. According to their analysis, increasing the prices of unhealthy foods such as processed meats and sugary sodas by 10%, while reducing the prices of healthy foods such as fruit and vegetables by 10%, would prevent an estimated 3.4% of yearly deaths in the U.S. Changing prices by 30% would have an even stronger effect, preventing an estimated 9.2% of yearly deaths. This data comports with that found in other countries, such as "previous modeling studies in South Africa and India, where a 20% SSB [sugar-sweetened beverage] tax was estimated to reduce diabetes prevalence by 4% over 20 years." The effects of price adjustments would be most pronounced on persons of lower socioeconomic status, as the researchers "found an overall 18.2% higher price-responsiveness for low versus high SES [socioeconomic status] groups."

This differential effect based on socioeconomic status contributes to concerns about such interventions, however. In *Harvard Public Health Review*, Kates and Hayward ask: "Well intentioned though they may be, at what point do these taxes overstep government influence on an individual's right to autonomy in decision-making? On whom does the increased financial burden of this taxation fall?" They note that taxes on sugar-sweetened beverages, for instance, "are likely to have a greater impact on low-income individuals…because individuals in those settings are more likely to be beholden to cost when making decisions about food."

However, "well intentioned though they may be," the worries that Kates and Hayward express are to some extent misguided. In particular, the idea that taxing unhealthy foods would burden those least able to afford it misses the point. Although the increased taxes would affect anyone who continued to purchase the items despite the higher prices, the goal of raising prices on unhealthy foods is precisely to dissuade people from buying them. As Kates and Hayward themselves remark, "Those in low-income environments may also be the largest consumers of obesogenic foods and therefore most likely to benefit from such a lifestyle change indirectly posed by SSB taxes." As the goal of the taxes is to promote those lifestyle changes, the financial burden objection is a non-starter.

Given this recognition, the question regarding autonomy constitutes a more substantial issue. Nevertheless, that concern also rests on a dubious assumption, as people's autonomy is not necessarily respected in the current situation either. The fact that those of lower socioeconomic status are more likely to have poorer diets suggests that such persons' food choices are the result of financial constraint, not fully autonomous, rational deliberation. Hence, by making healthy foods more affordable relative to unhealthy ones, government intervention might actually facilitate autonomous choices rather than hindering them.

On the other hand, suppose that the disproportionate consumption of unhealthy foods—and associated higher incidence of disease—among certain groups is not the result of financial hardship but rather the result of those persons' perceived self-interest. If so, that would suggest that members of these groups are being encouraged to persist in harmful dietary habits for the sake of corporate profits. In that case, violating autonomy for the sake of health may be permissible, as that practice would be morally preferable to the present system of corporate exploitation.

*Food Costs and Disease* ©UWorld

Annotations for this passage can be found in the CARS Passage Booklet.

## 1c. Direct Passage Claims Question 2

The author argues that concern about the financial burden of taxes on unhealthy foods is misguided because:

A. that burden is a less substantial issue than questions regarding autonomy.
B. the people most affected by price increases would also be most able to afford them.
C. the increased cost of those foods is designed to prevent people from buying them.
D. the added expense of buying some foods would be offset by the reduced prices on others.

See next page for the *strategy-based explanation* of this question.

> ## 1c. Direct Passage Claims Question 2
> ### Strategy-based Explanation
>
> The author argues that concern about the financial burden of taxes on unhealthy foods is misguided because:
>
> A. that burden is a less substantial issue than questions regarding autonomy.
> B. the people most affected by price increases would also be most able to afford them.
> C. the increased cost of those foods is designed to prevent people from buying them.
> D. the added expense of buying some foods would be offset by the reduced prices on others.

Looking at this question, we can see that most of it is telling us something rather than asking us something. Specifically, it describes a conclusion that the author argues for, that "concern about the financial burden of taxes on unhealthy foods is misguided." All that the question actually asks is the reason the author gives for this conclusion. Therefore, we just need to review the author's claims to see what this reason is.

## Applying the Method

*Passage Excerpt*

**[P4]** However, "well intentioned though they may be," the worries that Kates and Hayward express are to some extent misguided. In particular, the idea that taxing unhealthy foods would burden those least able to afford it **misses the point**. Although the increased taxes would affect anyone who continued to purchase the items despite the higher prices, **the goal of raising prices on unhealthy foods is precisely to dissuade people from buying them**.

### Step 1: Verify What the Passage Directly States

In Paragraph 4, we find passage text that closely resembles the wording of the question: the word "misguided" appears in both places, and "financial burden" is a shorter version of "burden those least able to afford it." So, why does the author say the concern about this burden is misguided?

They first claim that the concern "**misses the point**," and the next sentence provides a more detailed elaboration: "**the goal of raising prices on unhealthy foods is precisely to dissuade people from buying them**." Accordingly, this is the author's direct claim about why the concern is misguided.

### Step 2: Compare Answer Choices with Those Statements

We have verified the author's direct claim in the passage: "**the goal of raising prices on unhealthy foods is precisely to dissuade people from buying them**." Hence, all we need to do is find the answer choice that conveys the same information.

**Choice C** says: *the increased cost of those foods is designed to prevent people from buying them.*

This statement expresses the same point that the author makes in the passage, only in slightly different words. Therefore, **Choice C** is the correct answer.

**Another Look at the Answer Choices**

You might notice that Choice C is the only one that refers to the *point* of the food taxes, by mentioning what they are "designed" to do. This is an additional clue that we have used our method correctly, because we know the author claims that the concern about food taxes "misses the point." Thus, even if we weren't sure whether "dissuade" in the passage means the same thing as "prevent" in Choice C, we could still conclude that Choice C probably represents the direct claim made by the author.

For an alternative method of explanation based more specifically on passage evidence, you can view this question in the UWorld Qbank (sold separately).

---

**Relevant Subskill:**
**1c. Direct Passage Claims vs. 1e. Identifying Passage Perspectives**

Given that the question asks about something the author argues, you might wonder why it is classified under the **1c. Direct Passage Claims** subskill. Isn't this a question about the author's perspective? However, notice that this question does not actually ask you to identify the author's perspective; it identifies that perspective for you, and then asks you to identify a statement that the author makes in the passage. You can see this fact reflected in how we could answer the question simply by recognizing the relevant passage text.

Lesson 5.5 discusses **1e. Identifying Passage Perspectives** questions and how they can best be approached. In general, however, these kinds of observations help illustrate how the different subskills correspond to various question types and what is required to answer them.

## Passage H: For Whom the Bell Toils

In nineteenth-century America, most people dismissed the notion that someone might assassinate the president. The presumption was based not only on ethics but practicality: a president's term is inherently limited, and an unpopular one could be voted out of office. Therefore, it was reasoned, there would be no need to consider removal through violence. This belief persisted even after the shocking murder of Abraham Lincoln in 1865, which was viewed as an aberration. Thus it was that on July 2, 1881, Charles Guiteau could simply walk up to President James A. Garfield and shoot him in broad daylight. As Richard Menke portrays events, "Guiteau was in fact a madman who had come to identify with a disgruntled wing of the Republican Party after his deranged fantasies of winning a post from the new administration had come to nothing." Believing that God had told him to kill the president, Guiteau thought this act would garner fame for his religious ideas and thereby help to usher in the Apocalypse.

In an interesting parallel, Garfield had felt a sense of divine purpose for his own life after surviving a near-drowning as a young man. Unlike Guiteau's fanatical ravings, however, Garfield's vision worked to the betterment of himself and the world. Candice Millard describes his ascent from extreme poverty to incredible excellence in college, where "by his second year…they made him a professor of literature, mathematics, and ancient languages." Garfield would go on to join the Union Army, where he attained the rank of major general and argued that black soldiers should receive the same pay as their white compatriots. While serving in the war he was nominated for the House of Representatives but accepted the seat only after President Lincoln declared that the country had more need of him as a congressman than as a general. The reluctant politician would later himself become president under similar circumstances, after multiple factions of a deadlocked Republican convention unexpectedly nominated him instead of their original candidates in 1880. An honest man who opposed corruption within the party, Garfield strove both to heal the fractures of the Civil War and to uphold the aims for which it was fought, until "the equal sunlight of liberty shall shine upon every man, black or white, in the Union."

Although Guiteau's bullet would ultimately dim this light for Garfield, the president actually survived the initial attack and for a time appeared headed for recovery. Tragically, however, the hubris shown by his main physician, Dr. Willard Bliss, would lead instead to weeks of prolonged suffering. None of the doctors who examined Garfield were able to locate the bullet, and its lingering presence—along with the unwashed hands of the doctors who probed for it—led to an infection. As the president's condition worsened, inventor Alexander Graham Bell attempted to adapt his patented telephone technology to locate foreign metal in the human body. Inspired by speculation that the bullet's electromagnetic properties might be detectable, Bell used his newly developed "Induction Balance" device to listen for the sounds of electrical interference he hoped would isolate the site of the bullet.

Unfortunately, Bell's searches were unsuccessful. Like Garfield's doctors, he had been looking for the bullet in the wrong area. Menke asserts that Bell's efforts "would probably have fallen short" regardless. However, other historians suggest that Dr. Bliss, unwilling to consider challenges to his original assessment, prevented Bell from more thoroughly searching the president's body. Certainly, Bliss ignored the advice and protestations of other physicians, even as Garfield continued to decline. With death imminent, Garfield asked to be taken to his seaside cottage, where he died on the 19th of September.

*For Whom the Bell Toils* ©UWorld

Annotations for this passage can be found in the CARS Passage Booklet.

## 1c. Direct Passage Claims Question 3

Which of the following individuals took actions that contributed to President Garfield's eventual death from infection?

    I. Dr. Willard Bliss
    II. Alexander Graham Bell
    III. Charles Guiteau

A. III only
B. I and II only
C. I and III only
D. I, II, and III

See next page for the *strategy-based explanation* of this question.

Chapter 5: Skill 1 – Foundations of Comprehension

---

### 1c. Direct Passage Claims Question 3
### Strategy-based Explanation

Which of the following individuals took actions that contributed to President Garfield's eventual death from infection?

    I. Dr. Willard Bliss
    II. Alexander Graham Bell
    III. Charles Guiteau

A. III only
B. I and II only
C. I and III only
D. I, II, and III

---

The question asks whether particular individuals "*took actions that contributed to* President Garfield's eventual death from infection." This wording suggests we are looking for anyone whose actions played a role in causing Garfield to die from infection, regardless of whether that role was direct or indirect.

Since this is a Roman numeral question, we will need to consider what the passage says about each of the three individuals listed: **Dr. Willard Bliss, Alexander Graham Bell**, and **Charles Guiteau**. However, our overall method is the same; we just have to apply it multiple times.

## Applying the Method

*Passage Excerpt*

**[P1]** Thus it was that on July 2, 1881, **Charles Guiteau could simply walk up to President James A. Garfield and shoot him in broad daylight.**

**[P3]** Although **Guiteau's bullet would ultimately dim this light for Garfield**, the president actually survived the initial attack and for a time appeared headed for recovery. Tragically, however, **the hubris shown by his main physician, Dr. Willard Bliss, would lead instead to weeks of prolonged suffering.** None of the doctors who examined Garfield were able to locate **the bullet, and its lingering presence—along with the unwashed hands of the doctors who probed for it—led to an infection.** As the president's condition worsened, **inventor Alexander Graham Bell attempted to adapt his patented telephone technology to locate foreign metal in the human**

### Step 1: Verify What the Passage Directly States

**Option III: Charles Guiteau**

**Charles Guiteau is the person who shot Garfield**, and **the presence of the bullet** was one cause that **led to the infection**. Therefore, Option III should be part of the correct answer.

**Option I: Dr. Willard Bliss**

**Dr. Willard Bliss** is identified as **Garfield's main physician**, so he was probably one of the **doctors who probed for the bullet with unwashed hands**. In addition, the passage seems to blame Dr. Bliss for Garfield's "**prolonged suffering**," stating that **he ignored other physicians' recommendations even though Garfield continued to decline**. So, Option I should be part of the correct answer.

body. Inspired by speculation that the bullet's electromagnetic properties might be detectable, **Bell used his newly developed "Induction Balance" device to listen for the sounds of electrical interference he hoped would isolate the site of the bullet**.

**[P4] Unfortunately, Bell's searches were unsuccessful**. Like Garfield's doctors, he had been looking for the bullet in the wrong area. Menke asserts that Bell's efforts "would probably have fallen short" regardless. However, **other historians suggest that Dr. Bliss, unwilling to consider challenges to his original assessment, prevented Bell from more thoroughly searching the president's body. Certainly, Bliss ignored the advice and protestations of other physicians, even as Garfield continued to decline**. With death imminent, Garfield asked to be taken to his seaside cottage, where he died on the 19th of September.

**Option II: Alexander Graham Bell**

**Alexander Graham Bell tried to use his technology to find the bullet, but he was unsuccessful**. However, being unsuccessful in helping is not the same thing as causing harm. Moreover, the passage says nothing that connects Bell's actions to either the **presence of the bullet** or **doctors' unwashed hands**. Thus, Option II should *not* be part of the correct answer.

### Step 2: Compare Answer Choices with Those Statements

Based on our review of the passage, the actions of both **Charles Guiteau (Option III)** and **Dr. Willard Bliss (Option I)** contributed to Garfield's death from infection. However, **Alexander Graham Bell (Option II)** is not connected to either of the infection's causes.

Therefore, the correct answer should be *I and III only*. Looking at the answer choices, we see that this combination appears as Choice C. Accordingly, **Choice C** is the correct answer.

For an alternative method of explanation based more specifically on passage evidence, you can view this question in the UWorld Qbank (sold separately).

# Passage B: Jackie in 500 Words

Born in New York in 1929, Jacqueline Bouvier first came into the public eye as the wife of the 35th president of the United States, John F. Kennedy.  The president was assassinated in 1963, and by the end of what turned out to be a turbulent decade, Mrs. Kennedy had transformed herself into the enigmatic Jackie O., wife of Greek shipping magnate Aristotle Onassis.  Multifaceted and always elusive, the former first lady never ceased to fascinate; however, people had to be satisfied with only glimpses of this fashion icon, culture advocate, historic preservationist, polyglot, equestrienne, and book editor.  Indeed, upon her death in 1994, Jacqueline Kennedy Onassis was described as "the most intriguing woman in the world."  Often topping lists of the most admired individuals of the second half of the 20th century, this celebrated woman is likely someone many wish they had known.  Barring such a possibility, the best way to fully appreciate Jackie's exceptional nature might be to consider the people she wished she had known.

It is unsurprising that this woman who captured the public's imagination for decades distinguished herself from her peers early on.  Notably, in 1951, Ms. Bouvier entered a scholarship contest sponsored by *Vogue* and open to young women in their final undergraduate year, the annual Prix de Paris.  Among other assignments, applicants were asked to compose a 500-word essay, "People I Wish I Had Known," spotlighting three individuals influential in art, literature, or culture.  The future first lady chose an iconoclastic trio from the Victorian era: the French symbolist poet Charles Baudelaire, the Irish wit Oscar Wilde, and the innovative Ballets Russes dance company founder Sergei Diaghilev.

In a brief composition, Jackie provided deep insights into this bohemian threesome of poet, aesthete, and impresario with whom she strongly identified.  She concluded that Baudelaire deployed "venom and despair" as "weapons" in his poetry.  She idolized Wilde for being able "with the flash of an epigram to bring about what serious reformers had for years been trying to accomplish."  Diaghilev she defined as an artist of a different sort, someone who "possessed what is rarer than artistic genius in any one field—the sensitivity to take the best of each man and incorporate it into a masterpiece."  As Jackie poignantly observed, such a work is "all the more precious because it lives only in the minds of those who have seen it," dissipating soon after.  Furthermore, although these men espoused different disciplines, she discerned that "a common theory runs through their work, a certain concept of the interrelation of the arts."  Finally, foreshadowing her self-assumed role in the White House as the nation's unofficial minister of the arts, Jackie paid homage with her vision: "If I could be a sort of Overall Art Director of the Twentieth Century, watching everything from a chair hanging in space, it is their theories of art that I would apply to my period."

The contest committee judged Jackie's essay to have exhibited a profound appreciation for the arts combined with a truly outstanding level of intellectual maturity and originality of thought.  Similarly, biographer Donald Spoto deemed Jackie "remarkably unorthodox," not unlike the men about whom she wrote in her unusual composition, which he pronounced "a masterpiece of perceptive improvisation."  Thus, from a pool of 1,279 applicants representing 224 colleges, Jacqueline Bouvier was declared the winner.

Although Ms. Bouvier went on to decline the prestigious award, which would have involved living and working in Paris, she never gave up her dream of being the century's art director.  As first lady, she tirelessly promoted the arts and culture.  Today, the John F. Kennedy Center for the Performing Arts in Washington, DC, is a legacy of Jackie's vision.

*Jackie in 500 Words* ©UWorld

Annotations for this passage can be found in the CARS Passage Booklet.

### 1c. Direct Passage Claims Question 4

Based on the passage, when evaluating essays for the Prix de Paris scholarship contest, the committee seemed to consider:

　I. the applicant's level of intellectual maturity.
　II. the applicant's appreciation for the arts.
　III. the degree of original thought displayed by the applicant's essay.

- A. II only
- B. III only
- C. I and II only
- D. I, II, and III

See next page for the *strategy-based explanation* of this question.

> ### 1c. Direct Passage Claims Question 4
> ### Strategy-based Explanation
>
> Based on the passage, when evaluating essays for the Prix de Paris scholarship contest, the committee seemed to consider:
>
>     I. the applicant's level of intellectual maturity.
>     II. the applicant's appreciation for the arts.
>     III. the degree of original thought displayed by the applicant's essay.
>
>   A. II only
>   B. III only
>   C. I and II only
>   D. I, II, and III

This question asks you to consider how the Prix de Paris contest committee evaluated their scholarship essays. It is also a Roman numeral question, listing three possible criteria for this evaluation: **the applicant's level of intellectual maturity**, **the applicant's appreciation for the arts**, and **the degree of original thought displayed by the applicant's essay**. Fortunately, however, the discussion of how the committee judged Jackie's winning submission is confined to Paragraph 4. Therefore, these three options can be easily compared with the relevant information.

## Applying the Method

*Passage Excerpt*

[P4] The contest committee judged Jackie's essay to have exhibited **a profound appreciation for the arts** combined with **a truly outstanding level of intellectual maturity** and **originality of thought**. Similarly, biographer Donald Spoto deemed Jackie "remarkably unorthodox," not unlike the men about whom she wrote in her unusual composition, which he pronounced "a masterpiece of perceptive improvisation." Thus, from a pool of 1,279 applicants representing 224 colleges, Jacqueline Bouvier was declared the winner.

**Step 1: Verify What the Passage Directly States**

Option II: the applicant's appreciation for the arts.

The contest committee judged Jackie's essay as having exhibited a **profound appreciation for the arts**. Thus, Option II should be part of the correct answer.

Option I: the applicant's level of intellectual maturity.

The committee also thought Jackie's essay showed **a truly outstanding level of intellectual maturity**. Accordingly, Option I should be part of the correct answer.

Option III: the applicant's originality of thought.

Finally, the committee also believed Jackie to have displayed outstanding **originality of thought** in her essay. Therefore, Option III should be part of the correct answer.

## Step 2: Compare Answer Choices with Those Statements

After reviewing the direct passage claims in Paragraph 4, we concluded that **the applicant's appreciation for the arts (Option II)**, **the applicant's level of intellectual maturity (Option I)**, and **the degree of original thought displayed by the applicant's essay (Option III)** all constitute criteria by which the contest committee evaluated Jackie's winning scholarship essay.

Accordingly, the correct answer should be *I, II, and III*. Looking at the answer choices, we see that **Choice D** includes all three options and hence is correct.


For an alternative method of explanation based more specifically on passage evidence, you can view this question in the UWorld Qbank (sold separately).

## Passage I: Meaning: Readers or Authors?

Of late it has become popular among linguists and literary theorists to assert that a work's *meaning* depends upon the individual reader. It is readers, we are told, not authors, who create meaning, by interacting with a text rather than simply receiving it. Thus, a reader transcends the aims of the author, producing their own reading of the text. Indeed, on this line of thinking, even to speak of "the" text is to commit a conceptual error; every text is in fact many texts, a plurality of interpretations that resist comparative evaluation. This view is nonsense. That many otherwise sensible scholars should be attracted to it can perhaps be readily explained, but we should first delineate why the theory goes so far astray....

The absurdity of the view can be demonstrated by a practical analogy. Suppose Smith is conveying his ideas to Jones in conversation (the particular topic is of no consequence). Afterward, we discover that the men differ in their accounts of what Smith had expressed. At this point, Jones may decide that he misunderstood Smith, or perhaps that Smith was unclear. A more complex supposition might be that Smith misused some key term, so his words did not fully match his intentions. Any of these possibilities would reasonably describe why Jones and Smith possessed different opinions about what Smith had said.

What Jones may *not* justifiably conclude is that his own interpretation is what Smith *really meant*. He may not, in effect, say: "Yes, I admit that Smith honestly claims to have been saying something different, but I have formed my own equally correct understanding." Someone who made such an assertion would be suspected of making a joke; if he proved to be serious, we could only conclude that he was deeply confused or else being deliberately quarrelsome. For in questions about what Smith meant, it is surely Smith whose answer must be accepted…. [T]his is not a matter of *agreeing* with a speaker; Jones might judge Smith's ideas to be wrong, unfounded, etc. But whether Smith's ideas are right or wrong is a different matter from what those ideas *are*. On that count, Smith must be the authority.

However, this observation is in no way changed if Smith's ideas are written rather than spoken—sent by letter, for instance. Regardless of any interpretation Jones may produce, the letter's true meaning is whatever Smith intended to convey. Likewise it is, then, with a book, poem, or whatsoever object of literature a scholar (or ordinary reader) encounters. The writing down of ideas does not magically imbue them with malleability or render their content amorphous. From the loftiest tomes of Shakespeare or Milton to the lowliest of yellowed paperbacks, authors produce works with a particular message in mind. It is readers' task to discern that message, not to superimpose their own volitional perspectives.

To think otherwise is to undermine the foundation of literary scholarship. For what is the purpose of such scholarship, if not to seek understanding of an author's creation? One examines the text, taking note of style, historical context, allusions to other works, and other factors, in addition, of course, to the surface sense of the words themselves. If such an enterprise is to be reasonable, it must presume the existence of standards for success: accuracy and inaccuracy, depth or shallowness of analysis, grounds for preferring one interpretation to another. Different readers may come to different conclusions about a text, it is true. But to excise authorial intent from the evaluation of those conclusions does a disservice both to individual works and to literary study as a discipline.

*Meaning: Readers or Authors?* ©UWorld

Annotations for this passage can be found in the CARS Passage Booklet.

## 1c. Direct Passage Claims Question 5

Which of the following distinctions between related concepts is NOT made by the passage author?

A. Greater and lesser types of literature
B. An idea's content and its correctness
C. Credible and non-credible scholars
D. Accurate and inaccurate interpretations

See next page for the *strategy-based explanation* of this question.

> ### 1c. Direct Passage Claims Question 5
> ### Strategy-based Explanation
>
> Which of the following distinctions between related concepts is NOT made by the passage author?
>
> A. Greater and lesser types of literature
> B. An idea's content and its correctness
> C. Credible and non-credible scholars
> D. Accurate and inaccurate interpretations

This question includes two complicating factors. First, it asks what the author does NOT say, which means we will have to review more of the passage than usual to verify all the relevant passage claims. Second, it mentions no specific topics to look for, so we cannot take our typical approach of identifying what the answer should be before we look at the answer choices. Although that approach is beneficial for answering most questions, it is not possible with the way this question is phrased.

Nevertheless, we do know that we are looking for "distinctions between related concepts" and that *three* such distinctions must appear in the passage. Therefore, the correct answer will either include concepts that are not discussed in the passage at all, or concepts that the author does not draw a distinction between.

Turning to the answer choices, we need to find where the passage discusses the following topics: **literature** (greater vs. lesser); **ideas** (content vs. correctness); **scholars** (credible vs. non-credible); and **interpretations** (accurate vs. inaccurate).

## Applying the Method

*Passage Excerpt*

[P1] Indeed, on this line of thinking, even to speak of "the" text is to commit a conceptual error; every text is in fact many texts, a plurality of interpretations that resist comparative evaluation. This view is nonsense. That **many otherwise sensible scholars** should be attracted to it can perhaps be readily explained, but we should first delineate why the theory goes so far astray....

**Step 1: Verify What the Passage Directly States**

Choice C: Scholars

Paragraph 1 refers to "**otherwise sensible scholars**," or scholars who are generally sensible but wrong about a particular idea. However, the author does not seem to describe any scholars as non-sensible.

Paragraph 4 draws a distinction between **scholars** and **ordinary readers**, but not between two types of scholars.

**[P3]** [T]his is not a matter of *agreeing* with a speaker; Jones might judge Smith's ideas to be wrong, unfounded, etc. But **whether Smith's ideas are right or wrong is a different matter from what those ideas *are*.** On that count, Smith must be the authority.

**Choice B: Ideas**

Paragraph 3 contrasts **whether a person's ideas are right or wrong** (whether they are correct) from **what those ideas are** (their content).

**[P4]** Likewise it is, then, with a book, poem, or whatsoever **object of literature a scholar (or ordinary reader)** encounters. The writing down of ideas does not magically imbue them with malleability or render their content amorphous. **From the loftiest tomes of Shakespeare or Milton to the lowliest of yellowed paperbacks,** authors produce works with a particular message in mind. It is readers' task to discern that message, not to superimpose their own volitional perspectives.

**Choice A: Literature**

Paragraph 4 refers to **objects of literature**, and then mentions **the loftiest tomes of Shakespeare or Milton** and **the lowliest of yellowed paperbacks** as examples.

**[P5]** If such an enterprise is to be reasonable, it must presume the existence of standards for success: **accuracy and inaccuracy**, depth or shallowness of analysis, **grounds for preferring one interpretation to another**.

**Choice D: Interpretations**

Paragraph 5 lists **accuracy and inaccuracy** as factors involved in **judging interpretations**.

---

### Step 2: Compare Answer Choices with Those Statements

Having reviewed the relevant passage claims, we see that the author does discuss all four of the *topics* in the answer choices: **literature, ideas, scholars,** and **interpretations**. However, not all of these topics include a distinction between related concepts. While the author does mention scholars, the only description they give of these scholars is "otherwise sensible." Thus, the author does not seem to distinguish between *credible and non-credible scholars*, suggesting that **Choice C** is the correct answer.

The only other answer choice that might make us hesitate is Choice A, as the passage never uses the phrase *greater and lesser types of literature*. However, the author does draw a distinction between these types of literature, contrasting "the loftiest tomes" with "the lowliest of yellowed paperbacks." Therefore, we can rule out Choice A and be confident that **Choice C** is correct.

For an alternative method of explanation based more specifically on passage evidence, you can view this question in the UWorld Qbank (sold separately).

Lesson 5.4

# CMP Subskill 1d. Implicit Claims or Assumptions

| Skill 1: Foundations of Comprehension ||
|---|---|
| **Subskill** | **Student Objective** |
| 1a. Main Idea or Purpose | Identify the central theme or purpose of the passage |
| 1b. Meaning of Term | Determine the meaning of a term as used in the passage |
| 1c. Direct Passage Claims | Identify claims made directly in the passage |
| **1d. Implicit Claims or Assumptions** | **Identify implicit claims or unstated assumptions in the passage** |
| 1e. Identifying Passage Perspectives | Identify perspectives or positions presented in the passage |
| 1f. Further Implications of Passage Claims | Recognize further implications of claims stated in the passage |

**1d. Implicit Claims or Assumptions** questions ask you to identify ideas that are suggested by passage information but not directly expressed. For instance, a passage might include claims that imply a particular conclusion that is never overtly stated. Or, a point made by the author might presume some other statement is true, but this second statement is left unsaid. Thus, although the claims these questions ask about are "in" the passage, they are not conveyed explicitly.

## Method for Answering 1d. Implicit Claims or Assumptions Questions

### Step 1: Verify What the Passage Directly States

The first step in identifying an implicit claim is to review what is directly stated about the question topic. These statements provide the framework for what a source only assumes or hints at in the passage. As with other types of questions, this process also helps to ensure you haven't overlooked or misremembered any details.

### Step 2: Ask What Is Left Unsaid or Taken for Granted

Once you have verified the information in the relevant passage statements, ask yourself what these claims presume but don't actually say. What underlying points seem to be part of that passage information, even though they are not expressed? These implicit points will be closely related to the direct passage statements and will serve to clarify or further explain them.

### Step 3: Compare Answer Choices with the Implicit Claims

Finally, review the answer choices. You will find that one of them explicitly states something that is only suggested by the direct statements you considered in Steps 1 and 2. Accordingly, this answer represents the implicit claim or unstated assumption that you are looking for.

It is worth noting that identifying implicit claims is something we do all the time. When you speak with another person, you probably notice a number of things that they believe or take for granted but do not explicitly say. Thus, you already have the skill to recognize unstated claims; **1d. Implicit Claims or Assumptions** questions just ask you to apply that skill to CARS passages.

---

### What If the Correct Answer Is Explicitly Stated?

If a question asks what is assumed or implied, then you probably expect the correct answer to not be directly stated in the passage. However, the writers of the MCAT do not always craft questions as carefully as they should. Hence, you may come across a few cases where a question seems to be asking for something implicit yet the correct answer appears explicitly in the passage.

Although these situations are rare, they can be frustrating when they do occur. In dealing with them, it is a good idea to follow this principle:

When a question asks for an implicit claim, **do not rule out an answer choice *only* because it is directly stated**.

In other words, if an answer choice is directly stated instead of implicit, but it is **right in every *other* way**, then **it is probably the correct answer**. By keeping this principle in mind, you may still be annoyed at the question for being poorly phrased. However, you won't be annoyed at yourself for ruling out a correct answer you otherwise would have chosen.

## Subskill 1d. Implicit Claims or Assumptions
## (4 Practice Questions)

We now turn to four examples from the UWorld Qbank. In each case, the passage and question are first given without commentary, allowing you to practice applying the method yourself. Then, the passage and question are presented again, this time with annotations of the passage (see the CARS Passage Booklet) and a step-by-step explanation of the question using the described method.

## Passage B: Jackie in 500 Words

Born in New York in 1929, Jacqueline Bouvier first came into the public eye as the wife of the 35th president of the United States, John F. Kennedy. The president was assassinated in 1963, and by the end of what turned out to be a turbulent decade, Mrs. Kennedy had transformed herself into the enigmatic Jackie O., wife of Greek shipping magnate Aristotle Onassis. Multifaceted and always elusive, the former first lady never ceased to fascinate; however, people had to be satisfied with only glimpses of this fashion icon, culture advocate, historic preservationist, polyglot, equestrienne, and book editor. Indeed, upon her death in 1994, Jacqueline Kennedy Onassis was described as "the most intriguing woman in the world." Often topping lists of the most admired individuals of the second half of the 20th century, this celebrated woman is likely someone many wish they had known. Barring such a possibility, the best way to fully appreciate Jackie's exceptional nature might be to consider the people she wished she had known.

It is unsurprising that this woman who captured the public's imagination for decades distinguished herself from her peers early on. Notably, in 1951, Ms. Bouvier entered a scholarship contest sponsored by *Vogue* and open to young women in their final undergraduate year, the annual Prix de Paris. Among other assignments, applicants were asked to compose a 500-word essay, "People I Wish I Had Known," spotlighting three individuals influential in art, literature, or culture. The future first lady chose an iconoclastic trio from the Victorian era: the French symbolist poet Charles Baudelaire, the Irish wit Oscar Wilde, and the innovative Ballets Russes dance company founder Sergei Diaghilev.

In a brief composition, Jackie provided deep insights into this bohemian threesome of poet, aesthete, and impresario with whom she strongly identified. She concluded that Baudelaire deployed "venom and despair" as "weapons" in his poetry. She idolized Wilde for being able "with the flash of an epigram to bring about what serious reformers had for years been trying to accomplish." Diaghilev she defined as an artist of a different sort, someone who "possessed what is rarer than artistic genius in any one field—the sensitivity to take the best of each man and incorporate it into a masterpiece." As Jackie poignantly observed, such a work is "all the more precious because it lives only in the minds of those who have seen it," dissipating soon after. Furthermore, although these men espoused different disciplines, she discerned that "a common theory runs through their work, a certain concept of the interrelation of the arts." Finally, foreshadowing her self-assumed role in the White House as the nation's unofficial minister of the arts, Jackie paid homage with her vision: "If I could be a sort of Overall Art Director of the Twentieth Century, watching everything from a chair hanging in space, it is their theories of art that I would apply to my period."

The contest committee judged Jackie's essay to have exhibited a profound appreciation for the arts combined with a truly outstanding level of intellectual maturity and originality of thought. Similarly, biographer Donald Spoto deemed Jackie "remarkably unorthodox," not unlike the men about whom she wrote in her unusual composition, which he pronounced "a masterpiece of perceptive improvisation." Thus, from a pool of 1,279 applicants representing 224 colleges, Jacqueline Bouvier was declared the winner.

Although Ms. Bouvier went on to decline the prestigious award, which would have involved living and working in Paris, she never gave up her dream of being the century's art director. As first lady, she tirelessly promoted the arts and culture. Today, the John F. Kennedy Center for the Performing Arts in Washington, DC, is a legacy of Jackie's vision.

*Jackie in 500 Words* ©UWorld

Annotations for this passage can be found in the CARS Passage Booklet.

### 1d. Implicit Claims or Assumptions Question 1

The passage suggests that Jackie's admiration for the individuals discussed in her essay was related to her:

- A. dislike for the artists of her own day.
- B. passion for the social causes highlighted in their works.
- C. appreciation for a holistic approach to artistic expression.
- D. emphasis on the actual artworks rather than artistic principles.

See next page for the *strategy-based explanation* of this question.

Chapter 5: Skill 1 – Foundations of Comprehension

> ### 1d. Implicit Claims or Assumptions Question 1
> ### Strategy-based Explanation
>
> The passage suggests that Jackie's admiration for the individuals discussed in her essay was related to her:
>
> A. dislike for the artists of her own day.
> B. passion for the social causes highlighted in their works.
> C. appreciation for a holistic approach to artistic expression.
> D. emphasis on the actual artworks rather than artistic principles.

This question essentially asks why Jackie admired the individuals she wrote about in her essay. Thus, we are most likely looking for something these individuals held in common that would explain this admiration.

## Applying the Method

*Passage Excerpt*

**[P3]** In a brief composition, Jackie provided deep insights into the bohemian threesome of poet, aesthete, and impresario **with whom she strongly identified**. She concluded that Baudelaire deployed "venom and despair" as "weapons" in his poetry. She idolized Wilde for being able "with the flash of an epigram to bring about what serious reformers had for years been trying to accomplish." Diaghilev she defined as an artist of a different sort, someone who "possessed what is rarer than artistic genius in any one field—the sensitivity to take the best of each man and incorporate it into a masterpiece." As Jackie poignantly observed, such a work is "all the more precious because it lives only in the minds of those who have seen it," dissipating soon after. Furthermore, although these men espoused different disciplines, she discerned that "**a common theory** runs through their work, **a certain concept of the interrelation of the arts**." Finally, foreshadowing her self-assumed role in the White House as the nation's unofficial minister of the arts, Jackie paid homage with her vision: "If I could be a sort of Overall Art Director of the Twentieth Century, watching everything from a chair hanging in

**Step 1: Verify What the Passage Directly States**

Much of Paragraph 3 describes reasons why Jackie was drawn to each of her essay subjects *individually*. However, the author makes the following claims about Jackie's attitude toward the three men as a group:

1. Jackie **strongly identified** with the men **about whom she wrote** in her scholarship essay.

2. Jackie believed these individuals shared **a common theory**, namely **"a certain concept of the arts as interrelated."**

3. Jackie said she **would apply this theory herself** if she could become an art director.

The first and third claims give no information about these men or their work; only the second statement describes specifically what Jackie thought they had in common.

space, **it is their theories of art that I would apply** to my period.

### Step 2: Ask What Is Left Unsaid or Taken for Granted

As we have seen, the passage makes only one claim about Jackie's view of her essay subjects collectively: **All three men viewed the arts as interrelated**. Accordingly, this claim is the only piece of passage information that could explain *why* Jackie identified with these men and wanted to apply their theories.

Therefore, we can conclude that **Jackie admired these men at least partly for their view of the arts as interrelated**. This reason for her admiration is never directly stated, but it seems to be taken for granted.

### Step 3: Compare Answer Choices with the Implicit Claims

Looking at the answer choices, we notice that **Choice C** refers to Jackie's *appreciation for a holistic approach to artistic expression*, or the value she placed on treating various types of arts as interrelated. This answer matches the view we identified as motivating Jackie's admiration for the individuals she wrote about. Accordingly, **Choice C** is correct.

For an alternative method of explanation based more specifically on passage evidence, you can view this question in the UWorld Qbank (sold separately).

## Passage E: The Divine Sign of Socrates

From his bare feet to his bald pate, the potentially shapeshifting figure of Socrates found in the literary tradition that arose after his controversial trial and death presents an intriguing array of oddities and unorthodoxies. Most conspicuously, his unshod and shabby sartorial state flaunted poverty at a time when the city of Athens had become obsessed with wealth and its trappings. Yet the philosopher's peculiar appearance was but a hint of the strange new calling he embraced. Inspired perhaps by the famous Delphic dictum "Know thyself," he embarked on a mission devoted to finding truth through dialogue. In what struck some as a dangerous new method of inquiry, he subjected nearly everyone he encountered to intense cross-examination, mercilessly exposing the ignorance of his interlocutors. Moreover, in a culture that still put stock in magic, the highly charismatic, entertaining, and at times infuriating Socrates appeared to be a sorcerer bewitching the aristocratic young men of Athens who followed him fanatically about the agora.

By all credible accounts, this exceedingly eccentric, self-styled radical truth-seeker had more than a whiff of the uncanny about him. As Socrates himself explains in Plato's *Republic*, he was both blessed and burdened with a supernatural phenomenon in the form of a *daimonion* or inner spirit that always guided him: "This began when I was a child. It is a voice, and whenever it speaks, it turns me away from something I am about to do, but it never encourages me to do anything." An overtly rational thinker, Socrates nonetheless considered these warnings—or, in James Miller's words, "the audible interdictions he experienced as irresistible"—to be infallible. Such oracular injunctions were highly anomalous as tutelary spirits were thought to assume a more nuanced presence. Some scholars have dismissed Socrates' recurring sign as a hallucination or psychological aberration. Others have conjectured that the internal voice might be attributable to the cataleptic or trancelike episodes from which the philosopher purportedly suffered. Indeed, as Miller notes, "Socrates was storied for the abstracted states that overtook him"; not infrequently, his companions would see him stop in his tracks and stand still for hours, completely lost in thought.

As Socrates further insisted, it was only the protestations of this apotreptic voice that held him back from entering the political arena. Even so, its personal admonitions could not spare him persecution. Despite the political amnesty extended by the resurgent democracy that succeeded the interim pro-Spartan oligarchy, the thinker's notoriety and ambiguous allegiances aroused suspicions. In 399 BCE, Socrates was brought before the court on trumped-up charges of impiety; these included willfully neglecting the traditional divinities, flagrantly introducing new gods to the city, and wittingly corrupting the youth. Athenian society recognized no division between religious and civic duties, and capricious gods demanded constant appeasement through sacrifices and rituals. Consequently, belief in a purely private deity—particularly a wholly benevolent deity conveying unequivocal messages—was inadmissible. Worse, as Socrates' own testimony revealed, he honored this personal god's authority above even the laws of the city. Hence, the philosopher's *daimonion* loomed over his indictment, conviction, and sentencing.

Nevertheless, in his defense speech as reconstructed by Plato in the *Apology*, Socrates maintained confidence in the protective nature and prophetic powers of his inner monitor. He never questioned its affirmatory silence toward his predicament, remarking, "The divine faculty would surely have opposed me had I been going to evil and not to good." Thus, Socrates acknowledged that his *daimonion* had its reasons, however inscrutable. Variously described as malcontent and martyr, public nuisance and prophet, laughingstock and hero, the mercurial Athenian, like the sign that guided him, was difficult to fathom yet impossible to ignore.

*The Divine Sign of Socrates* ©UWorld

Annotations for this passage can be found in the CARS Passage Booklet.

## 1d. Implicit Claims or Assumptions Question 2

The passage suggests that the most problematic aspect of Socrates' guiding spirit was:

- A. its show of benevolence.
- B. its manifestation as a voice.
- C. its clarity as an oracle.
- D. its threat to society.

See next page for the *strategy-based explanation* of this question.

> ## 1d. Implicit Claims or Assumptions Question 2
> ### Strategy-based Explanation
>
> The passage suggests that the most problematic aspect of Socrates' guiding spirit was:
>
> A. its show of benevolence.
> B. its manifestation as a voice.
> C. its clarity as an oracle.
> D. its threat to society.

The question asks about the *most* problematic aspect of Socrates' guiding spirit. Accordingly, it is possible that multiple such problems are described in the passage, and you must compare their nature and severity. By reviewing what the passage states about the effects of the spirit on Socrates' life, we will likely find that some are more pronounced than others, pointing us to the correct answer.

## Applying the Method

*Passage Excerpt*

**[P2]** As Socrates himself explains in Plato's *Republic*, he was both blessed and burdened with a supernatural phenomenon in the form of a *daimonion* or inner spirit that always guided him: "This began when I was a child. It is a voice, and whenever it speaks, it turns me away from something I am about to do, but it never encourages me to do anything." An overtly rational thinker, Socrates nonetheless considered these warnings—or, in James Miller's words, "the audible interdictions he experienced as irresistible"—to be infallible. Such oracular injunctions were highly anomalous as tutelary spirits were thought to assume a more nuanced presence. Some scholars have dismissed Socrates' recurring sign as a hallucination or psychological aberration. Others have conjectured that the internal voice might be attributable to the cataleptic or trancelike episodes from which the philosopher purportedly suffered.

**[P3]** In 399 BCE, Socrates was brought before the court on trumped-up charges of impiety; these included willfully neglecting the traditional divinities, flagrantly introducing new gods to the city, and wittingly corrupting the youth. Athenian society recognized no division between religious and civic duties, and capricious gods demanded constant

**Step 1: Verify What the Passage Directly States**

Both Paragraphs 2 and 3 provide information about Socrates' guiding spirit.

Paragraph 2 describes two features of this spirit:

1. **It was seen as unusual** for manifesting as an unambiguous voice.

2. **Some scholars think** it may have been caused by a psychological issue.

Paragraph 3 then describes more specifically how Socrates' guiding spirit affected his society.

3. **It didn't fit with official Athenian religious and civic duties**, which were **thought of as essential to the city's well-being.**

appeasement through sacrifices and rituals. Consequently, belief in a purely private deity—particularly a wholly benevolent deity conveying unequivocal messages—was inadmissible. Worse, as Socrates' own testimony revealed, he honored this personal god's authority above even the laws of the city. Hence, the philosopher's *daimonion* loomed over his indictment, conviction, and sentencing.

**4. Socrates admitted that he would follow this voice over the city's own laws**, which contributed to his conviction at trial.

Statements 1 and 2 do not seem to suggest any problems caused by Socrates' guiding spirit. By contrast, 3 and 4 portray it as troublesome for both Socrates and his society.

### Step 2: Ask What Is Left Unsaid or Taken for Granted

We have determined that Paragraph 3 presents two ways in which Socrates' spirit caused problems. If we consider these claims together, we find that:

**Socrates was convicted of following his own personal spirit instead of the deities viewed as essential to the well-being of Athens**.

Accordingly, what is taken for granted in this description is that:

**Socrates' devotion to his guiding spirit was seen as harmful to the well-being of Athens**.

### Step 3: Compare Answer Choices with the Implicit Claims

Looking at the answer choices, we note that **Choice D**, *its threat to society*, corresponds closely to the conclusion we have drawn. If Socrates was viewed as harming Athens by following his guiding spirit, then the Athenians considered the spirit a threat to their society's welfare. Accordingly, the passage suggests this threat was the most problematic aspect of Socrates' guiding spirit, making **Choice D** the correct answer.

For an alternative method of explanation based more specifically on passage evidence, you can view this question in the UWorld Qbank (sold separately).

# Passage J: The Inkblots

For almost 100 years now, the psychological evaluation known as the Rorschach Inkblot Test has engendered much controversy, including skepticism about its value, questions about its scoring, and, especially, criticism of its interpretive methods as too subjective. Thus, the Rorschach test, which emerged from the same early twentieth-century zeitgeist that produced Einstein's physics, Freudian psychoanalysis, and abstract art, seems one of modernity's most misbegotten children. Destined never to be completely accepted or discredited, the test remains a perennial outlier in its field. Nevertheless, the inkblots' mystery and aesthetic appeal have caused them to be indelibly printed on our cultural fabric.

The now iconic inkblots were introduced to the world by Swiss psychiatrist Hermann Rorschach in his 1921 book *Psychodiagnostics*. As both director of the Herisau Asylum in Switzerland and an amateur artist, Rorschach was uniquely positioned to wed the new practice of psychoanalysis to the budding phenomenon of abstract art. For instance, reading Freud's work on dream symbolism prompted him to recall his childhood passion for a game based on inkblot art called *Klecksographie*. He was also cognizant that in a recently published dissertation, his colleague Szymon Hens had used inkblots to try to probe the imagination of research subjects; moreover, a few years earlier, the French psychologist and father of intelligence testing Alfred Binet had used them to measure creativity.

Motivated by these developments, the Herisau director decided to revisit that childhood pastime that had awakened his curiosity about how visual information is processed. In particular, he wondered why different people saw different things in the same image. Traditionally, psychoanalysts had relied on language for insights; however, as biographer Damion Searls reports, Rorschach's theories would exemplify the principle that "who we are is a matter less of what we say than of what we see." Indeed, through a process of perception termed pareidolia, the mind projects meaning onto images, detecting in them familiar objects or shapes. Consequently, what a person sees in an image reveals more about that person than about the image itself.

Rorschach experimented with countless inkblots, eventually selecting ten—five black on white, two also featuring some red, and three pastel-colored—to use with research subjects. For these perfectly symmetrical images—each of which he was said to have "meticulously designed to be as ambiguous and 'conflicted' as possible"—the primary question was always "What do *you* see?" Rorschach was especially careful to note how much attention individuals paid to various components of each inkblot (such as form, color, and a sense of movement) and whether they concentrated on details or the whole image. Having observed that his patients with schizophrenia gave distinctly different responses from the control group, Rorschach envisioned his experiment as a diagnostic tool for the disease. Nevertheless, he resisted the notion that its results could be used to assess personality. In fact, until his untimely death from a ruptured appendix in 1922, Rorschach referred to his project as an "interpretive form experiment" rather than a test. Ironically, however, by the 1960s, the Rorschach Inkblot Test was known chiefly as a personality assessment and had become the most frequently administered projective personality test in the US.

Rorschach's test has survived nearly incessant scrutiny, including a 2013 comprehensive study of all Rorschach test data and repeated revisions to its scoring, yet doubts about its validity and reliability persist. Much like the inkblots themselves—which tantalize us with the possibility of divulging the secrets of who we are and how we see the world—the test has (for better or worse) defied attempts to fix its meaning. Thus, what has been called "the twentieth century's most visionary synthesis of art and science" stands tempered by harsh criticism.

*The Inkblots* ©UWorld

Annotations for this passage can be found in the CARS Passage Booklet.

## 1d. Implicit Claims or Assumptions Question 3

The passage suggests which of following statements about the Rorschach test?

A. Its results often contradict the findings of other types of psychological tests.
B. Its interpretation may be more subject to bias than other psychological tests.
C. It tends to misrepresent the personality traits of the test-taker.
D. It should not be used as part of a personality assessment.

See next page for the *strategy-based explanation* of this question.

> **1d. Implicit Claims or Assumptions Question 3**
> **Strategy-based Explanation**
>
> The passage suggests which of following statements about the Rorschach test?
>
> A. Its results often contradict the findings of other types of psychological tests.
> B. Its interpretation may be more subject to bias than other psychological tests.
> C. It tends to misrepresent the personality traits of the test-taker.
> D. It should not be used as part of a personality assessment.

The only reference this question provides is the mention of the Rorschach test. Accordingly, we must turn to the answer choices for more information on what to review. These choices are focused on two main areas: psychological testing and personality. Hence, we will need to examine the passage author's claims about those topics to determine what they leave unsaid or take for granted.

## Applying the Method

*Passage Excerpt*

**[P1]** For almost 100 years now, the psychological evaluation known as the Rorschach Inkblot Test has engendered much controversy, including skepticism about its value, questions about its scoring, and, **especially, criticism of its interpretive methods as too subjective**.

**[P3]** …however, as biographer Damion Searls reports, Rorschach's theories would exemplify the principle that **"who we are is a matter less of what we say than of what we see."** Indeed, through a process of perception termed pareidolia, the mind projects meaning onto images, detecting in them familiar objects or shapes. Consequently, **what a person sees in an image reveals more about that person than about the image itself**.

**[P4]** Rorschach envisioned his experiment as a diagnostic tool for the disease. Nevertheless, **he resisted the notion that its results could be used to assess personality**. In fact, until his untimely death from a ruptured appendix in 1922, Rorschach referred to his project as an "interpretive form experiment" rather than a test. Ironically, however, by the 1960s, the Rorschach Inkblot Test was known chiefly as a personality assessment and

### Step 1: Verify What the Passage Directly States

Paragraph 1 provides information on psychological testing. More specifically, it states that the Rorschach test's **interpretive methods have been criticized as too subjective**.

Paragraphs 3 and 4 discuss the Rorschach test in relation to personality.

According to Paragraph 3, Rorschach believed that **aspects of an individual's personality were revealed by what they saw in images**. In addition, this view can be connected to the psychological phenomenon of pareidolia.

Paragraph 4 states that although **Rorschach resisted using "his interpretive form experiment" as a personality assessment**, "ironically," the Rorschach test became a major test of personality in the 1960s.

had become the most frequently administered projective personality test in the US.

### Step 2: Ask What Is Left Unsaid or Taken for Granted

The information in Paragraph 1 does not discuss any other types of tests besides the Rorschach Inkblot Test. However, if *that* test's methods have been criticized as "too subjective," what seems to be left unsaid is that *other* tests are not criticized in this way. In other words, the author suggests that **the Rorschach test is more subjective than other types of psychological testing**.

In Paragraph 3, the author connects Rorschach's theories to the phenomenon of pareidolia, then draws a conclusion about the importance of "what we see." Thus, the author seems to take for granted that **this aspect of Rorschach's theories is correct**.

Paragraph 4 conveys that the Rorschach test is now used in a way that Rorschach never intended. Hence, what is left unsaid is that **Rorschach probably would not approve of how his inkblot test came to be used**.

### Step 3: Compare Answer Choices with the Implicit Claims

Looking at the answer choices, we see that **Choice B, *Its interpretation may be more subject to bias than other psychological tests***, closely corresponds to what we identified as implicit in Paragraph 1. Specifically, if **the Rorschach test is more subjective than other types of psychological testing**, then its interpretation depends more on an individual's judgment—or, in other words, it is more likely to be influenced by that individual's biases. Hence, **Choice B** is probably correct.

The only other choice that might also seem to be implied is Choice D, that Rorschach's test *should not be used as part of a personality assessment*. However, although the author implies that Rorschach himself may not have approved of this use of the inkblots, that is not the same thing as the author suggesting that the inkblots *should not* be used to assess personality. Accordingly, this answer is not well supported, and we should stick with **Choice B**.

For an alternative method of explanation based more specifically on passage evidence, you can view this question in the UWorld Qbank (sold separately).

## Passage F: The Victorian Internet

Lasting roughly from 1820 to 1914, the Victorian Era is often defined by its many distinctive sociological conditions, including industrialization, urbanization, railroad travel, imperialism, territorial expansionism, and the frictions sparked by Darwinism and democratic reform. However, descriptors that may not come as readily to mind are "the Information Age" and "the Age of Communication"; nor would we likely associate this era with something called the "highway of thought," which emanated from the electric telegraph—an invention that, in retrospect, could be renamed "the Victorian Internet." Yet the rise of the electric telegraph in the mid-1800s constituted the greatest revolution in communication since the invention of the printing press in the fifteenth century and until the launch of the World Wide Web at the end of the twentieth century.

Historically, messages could be conveyed only as fast as a person could travel from one location to another. However, by the end of the eighteenth century, the Chappe brothers had constructed a rudimentary optical telegraph on a hilltop tower in France. Using the large, jointed arms attached to the roof, operators could form various configurations to communicate a message to a similar tower farther away, which would then relay the message to a third tower, and so on, in a long-distance chain. Nevertheless, even with telescopes, visibility severely hampered the efficacy of this semaphore system. After countless attempts by scientists to use electricity to transmit messages, innovators in both Britain and America, including the painter and polymath Samuel F.B. Morse, worked at harnessing electromagnetic forces to send communications via cable. But while in Britain the technology was initially reserved for railway signaling, in the U.S. it culminated in the transmission of a message along a 40-mile telegraphic wire from Washington, D.C., to Baltimore in 1844. Using a binary code of short and long electrical impulses, or "dots and dashes," Morse dispatched the words: "What hath God wrought?"

Within 20 years, telegraph cables crisscrossed the continental U.S. and much of Europe. The telegraph office became ubiquitous, and telegrams—often bouncing from one office to another like emails tossed from server to server—reached ever more remote destinations. Following some spectacular failures, a durable cable was stretched across the floor of the Atlantic Ocean, enabling Europe and the U.S. to exchange messages within minutes. In the 1870s, the British Empire connected London with outposts in India and Australia. While individual telegram speed remained relatively constant, an endless deluge of information began to pour through the wires. There was, as Tom Standage observes, "an irreversible acceleration in the pace of business life," reflecting how "telegraphy and commerce thrived in a virtuous circle." It was as though "rapid long-distance communication had effectively obliterated time and space," begetting the phenomenon known as "globalization."

A new type of skilled worker, the telegrapher, was born. He or she belonged to a vast, online community, whose semi-anonymous members shared a unique intermediary role as well as a language of dots and dashes—vaguely prefiguring the exchange of bits and bytes along modern computer networks. Like today's online communities, these telegraphers fostered their own subculture: jokes and anecdotes flew over the wires; some operators invented systems to play games; and occasional romances blossomed. Regrettably, hackers, scammers, and shady entrepreneurs also frequented the byways of this early internet.

Predictably, newer forms of the original telegraphic technology—first the telephone and, much later, the fax machine—eventually encroached. The last telegram departed from India in 2013, but the twenty-first century has arguably seen a revival of this communication form in the text message. As Standage claims, texting has not only "reincarnate[d] a defunct nineteenth-century technology" but reinforced "the democratization of telecommunications" inaugurated by the miraculous Victorian Internet.

*The Victorian Internet* ©UWorld

Annotations for this passage can be found in the CARS Passage Booklet.

## 1d. Implicit Claims or Assumptions Question 4

Which of the following statements is implied by passage information?

   A. The transatlantic telegraph system provided less reliable service than the domestic system.
   B. The telegraph can be considered an essentially American invention.
   C. The speed of the typical telegram was continually increasing.
   D. European markets became heavily dominated by telegraphic communication.

See next page for the *strategy-based explanation* of this question.

Chapter 5: Skill 1 – Foundations of Comprehension

---

### 1d. Implicit Claims or Assumptions Question 4
### Strategy-based Explanation

Which of the following statements is implied by passage information?

    A. The transatlantic telegraph system provided less reliable service than the domestic system.
    B. The telegraph can be considered an essentially American invention.
    C. The speed of the typical telegram was continually increasing.
    D. European markets became heavily dominated by telegraphic communication.

---

This question is complicated by the fact that it is completely general and provides no reference points on what information it covers. Accordingly, you will need to review multiple areas of the passage to determine what is implied. More specifically, you will have to look at what the passage says about the transatlantic telegraph system, the invention of the telegraph, telegram speed, and European markets.

## Applying the Method

*Passage Excerpt*

**Step 1: Verify What the Passage Directly States**

[P2] After countless attempts by scientists to use electricity to transmit messages, **innovators in both Britain and America**, including the painter and polymath Samuel F.B. Morse, **worked at harnessing electromagnetic forces to send communications via cable. But while in Britain the technology was initially reserved for railway signaling, in the U.S. it culminated in the transmission of a message** along a 40-mile telegraphic wire from Washington, D.C., to Baltimore in 1844. Using a binary code of short and long electrical impulses, or "dots and dashes," Morse dispatched the words: "What hath God wrought?"

Paragraph 2 details the telegraph's invention. It reveals that **both British and American inventors** worked at finding ways to send long-distance communications. However, the resulting telegraph technology **was initially used for different purposes** in Britain and America.

[P3] Within 20 years, **telegraph cables crisscrossed the continental U.S. and much of Europe.... Following some spectacular failures, a durable cable was stretched across the floor of the Atlantic Ocean, enabling Europe and the U.S. to exchange messages within minutes**. In the 1870s, the British Empire connected London with outposts in India and Australia. While individual telegram speed remained relatively constant, **an endless deluge of information began to pour through the wires.** There was, as Tom Standage observes, **"an irreversible acceleration in the pace of business life,"**

In Paragraph 3, we find information about the transatlantic telegraph system, telegram speed, and international markets. We can note that:

1. The telegraph system expanded across both Europe and the U.S., **leading to reliable transatlantic communication**.

2. Although the speed of telegrams remained the same, **the volume of long-distance communications increased dramatically**. As a result, **the pace of business increased**, which meant that **telegraphic communication and commerce thrived simultaneously**.

94

reflecting how **"telegraphy and commerce thrived in a virtuous circle."** It was as though "rapid long-distance communication had effectively obliterated time and space," begetting the phenomenon known as **"globalization."**

3. The new ease of communication transformed the world by leading to **globalization**.

### Step 2: Ask What Is Left Unsaid or Taken for Granted

The claims in Paragraph 2 indicate that the telegraph was not an exclusively American invention but was also a British one.

However, most of the relevant information for our question is found in Paragraph 3. There, we see that the telegraph had the following effects:

- Enabling rapid, reliable communication within and between America and Europe;

- Increasing the pace and volume of commerce, while mutually growing in response to that increased commerce;

- Transforming the world through globalization.

Accordingly, what seems to be taken for granted about these effects is that **the telegraph became essential to life, communication, and business in America and Europe**.

### Step 3: Compare Answer Choices with the Implicit Claims

From the information found in Paragraph 3, we concluded that **the telegraph became essential to life, communication, and business in America and Europe**.

Looking at the answer choices, we note that **Choice D, *European markets became heavily dominated by telegraphic communication***, represents one aspect of this idea. Specifically, it reflects the transformative role of the telegraph in facilitating business in Europe (and throughout the world). Accordingly, **Choice D** is the correct answer.

For an alternative method of explanation based more specifically on passage evidence, you can view this question in the UWorld Qbank (sold separately).

Lesson 5.5

# CMP Subskill 1e. Identifying Passage Perspectives

| Skill 1: Foundations of Comprehension ||
| --- | --- |
| **Subskill** | **Student Objective** |
| 1a. Main Idea or Purpose | Identify the central theme or purpose of the passage |
| 1b. Meaning of Term | Determine the meaning of a term as used in the passage |
| 1c. Direct Passage Claims | Identify claims made directly in the passage |
| 1d. Implicit Claims or Assumptions | Identify implicit claims or unstated assumptions in the passage |
| **1e. Identifying Passage Perspectives** | **Identify perspectives or positions presented in the passage** |
| 1f. Further Implications of Passage Claims | Recognize further implications of claims stated in the passage |

**1e. Identifying Passage Perspectives** questions ask you to identify what the author or another source in the passage believes or feels about a topic. These questions may take different forms, such as:

- With which of the following statements would the author most likely agree?
- Based on the passage, [a quoted source] believes that:
- The author's attitude toward the passage topic can best be described as:

However, these variations are all ways of asking about a **viewpoint**, and thus they are all based on the same types of passage evidence.

# Method for Answering 1e. Identifying Passage Perspectives Questions

## Step 1: Take Note of Viewpoint Indicators

Fortunately, most authors are not shy about telling you what they believe. The positions expressed in a passage are typically accompanied by one or more **Viewpoint Indicators**: words or phrases that either reveal perspectives or introduce information that reveals them. Examples of such indicators include:

- **positive, negative, or neutral descriptions**

    "Intriguingly"; "that depressing assumption"; "will undoubtedly be heard in the future"

- **points that intentionally reference a view**

    "it behooves us to re-evaluate our perspective"; "As Paul Strathern notes"; "In opposition to singular market ownership, Keynesian economics points to..."

- **claims that a source concludes or argues for**

    "Hence, whether the monster is literal or metaphorical, these seemingly divergent interpretations in fact exhibit a clear and striking coherence."

You can think of these indicators as "flags" that mark the various perspectives in a passage. Thus, you may find it helpful to highlight them as you read, making it easy to reference viewpoints when they come up in a question.

## Step 2: Identify the Source's Belief or Attitude

By paying attention to Viewpoint Indicators, you can quickly identify the perspective mentioned in the question. Discerning what a source thinks or feels is often just a matter of noticing the adjectives they use or the way they introduce a claim. Once you have recognized the source's viewpoint, you should see that one answer choice matches your observation.

### Subskill Connection: 1e / 2e / 3e

As mentioned in Lesson 4.4, there are three different subskills that deal with passage perspectives, each corresponding to one of the overarching CARS skills: **1e. Identifying Passage Perspectives** (Foundations of Comprehension); **2e. Determining Passage Perspectives** (Reasoning Within the Text); and **3e. Applying Passage Perspectives** (Reasoning Beyond the Text).

These three types of questions are closely related, but vary in terms of what is needed to answer them. For instance, **1e. Identifying Passage Perspectives** questions typically ask about views that can be recognized without additional reasoning. However, **2e. Determining Passage Perspectives** and **3e. Applying Passage Perspectives** questions may require you to consider a greater amount of passage information or follow extra logical steps. The similarities and differences between these subskills are further discussed in Lessons 6.5 and 7.5.

## Subskill 1e. Identify Passage Perspectives
## (4 Practice Questions)

We now look at four questions from the UWorld Qbank. In each case, the passage and question are first given without commentary, allowing you to practice applying the method yourself. Then, the passage and question are presented again, this time with annotations of the passage (see the CARS Passage Booklet) and a step-by-step explanation of the question using the described method.

## Passage C: Probability and The Universe

The idea of probability is frequently misunderstood, in large part because of a conceptual confusion between objective probability and subjective probability. The failure to make this distinction leads to an erroneous conflation of genuine possibility with what is in fact merely personal ignorance of outcome. An example will clarify.

A standard die is rolled on a table, but the outcome of the roll is concealed. Should an observer be asked the chance that a particular number was rolled—five, say—the natural response is 1/6. However, this answer is incorrect. To say there is a one-in-six chance that a five was rolled implies there is an equal chance that any of the other numbers were rolled. But there is no equal chance, because the roll has already occurred. Hence, the probability that the result of the roll is a five is either 100% or 0%, and the same is true for each of the other numbers.

It might be objected that such an analysis is an issue of semantics rather than a substantive claim. For declaring the probability to be 1/6 is merely an expression that, for all we know, any number from 1 through 6 might have been rolled. But the difference between *for all we know* and what *is* remains crucially important, because it forestalls a tendency to make scientific assertions from a perspective biased by human perception....

[T]hus, it should be clear that claims about the purpose of the universe rest on shaky ground. In particular, we must be wary of inferences drawn from juxtaposing the existence of intelligent life with the genuinely improbable cosmological conditions that make such life possible. Joseph Zycinski provides an instructive account of those conditions: "If twenty billion years ago the rate of expansion were minimally smaller, the universe would have entered a stage of contraction at a time when temperatures of thousands of degrees were prevalent and life could not have appeared. If the distribution of matter were more uniform the galaxies could not have formed. If it were less uniform, instead of the presently observed stars, black holes would have formed from the collapsing matter." In short, the existence of life (let alone intelligent life) depends upon the initial conditions of the universe having conformed to an extremely narrow range of possible values.

It is tempting to go beyond Zycinski's factual point to draw deep cosmological, teleological, and even theological conclusions. But no such conclusions follow. The reason why a life-sustaining universe exists is that if it did not, there would be no one to wonder why a life-sustaining universe exists. This fact is a function not of purpose, but of pre-requisite. For instance, suppose again that a five was rolled on a die. We now observe a five on its face, not because the five was "meant to be" but because one side of the die had to land face up. Even if we imagine that the die has not six but six *billion* sides, that analysis is unaffected. Yes, it was unlikely that any given value would be rolled, but *some* value had to be rolled. That "initial condition" then set the parameters for what kind of events could possibly follow; in this case, the observation of the five.

Similarly, the existence of intelligent beings is evidence that certain physical laws obtained in the universe. However, it is not evidence that those beings or laws were necessary or intended rather than essentially random. The subjective probability of that outcome is irrelevant, because in this universe the objective probability of its occurrence is 100%. Thus, the seemingly low initial chance that such a universe would exist is not in itself indicative of a purpose to its existence.

*Probability and The Universe* ©UWorld

Annotations for this passage can be found in the CARS Passage Booklet.

> **1e. Identifying Passage Perspectives Question 1**
>
> Given the information in the passage, with which of the following statements about the quote from Joseph Zycinski would the author be most likely to agree?
>
> A. It explains objective and subjective probability.
> B. It exaggerates the point in question.
> C. It entails an erroneous conclusion.
> D. It provides clarifying information.

See next page for the *strategy-based explanation* of this question.

> ## 1e. Identifying Passage Perspectives Question 1
> ### Strategy-based Explanation
>
> Given the information in the passage, with which of the following statements about the quote from Joseph Zycinski would the author be most likely to agree?
>
> - A. It explains objective and subjective probability.
> - B. It exaggerates the point in question.
> - C. It entails an erroneous conclusion.
> - D. It provides clarifying information.

We can find Zycinski's quote in Paragraph 4, where it is used to explain the narrow conditions that are necessary for life to exist. More importantly for answering this question, we can see the author's attitude toward the quote reflected in the way that they describe it in the passage.

## Applying the Method

*Passage Excerpt*

**[P4] Joseph Zycinski provides an instructive account** of those conditions: "If twenty billion years ago the rate of expansion were minimally smaller, the universe would have entered a stage of contraction at a time when temperatures of thousands of degrees were prevalent and life could not have appeared. If the distribution of matter were more uniform the galaxies could not have formed. If it were less uniform, instead of the presently observed stars, black holes would have formed from the collapsing matter."

**[P5]** It is tempting to go beyond **Zycinski's factual point** to draw deep cosmological, teleological, and even theological conclusions. But no such conclusions follow.

### Step 1: Take Note of Viewpoint Indicators

**Viewpoint Indicator**

The author introduces Zycinski's quote by calling it "**instructive**." Accordingly, they see Zycinski's quote as providing a **useful explanation**.

**Viewpoint Indicator**

After Zycinski's quote has been presented, the author refers back to it, this time describing it as a "**factual point**." Therefore, they view the information in Zycinski's quote as being **correct**.

## Step 2: Identify the Source's Belief or Attitude

Based on the viewpoint indicators "**instructive**" and "**factual**," we can see that the author views Zycinski's quote as **explanatory** and **correct**. Thus, **Choice D** immediately stands out: *It provides clarifying information* seems to fit the author's view of the quote exactly.

The only potential alternative would be Choice A, *It explains objective and subjective probability*, which also portrays Zycinski's quote as explanatory. However, if this answer choice were correct, then Choice D would *also* have to be correct. That is, if Zycinski's quote explained objective and subjective probability, then it would be providing clarifying information. The same is not true in reverse, however; the quote could provide clarifying information without specifically explaining objective and subjective probability. Therefore, by comparing these two answer choices, we see that **Choice D** is the better answer.

For an alternative method of explanation based more specifically on passage evidence, you can view this question in the UWorld Qbank (sold separately).

## Passage F: The Victorian Internet

Lasting roughly from 1820 to 1914, the Victorian Era is often defined by its many distinctive sociological conditions, including industrialization, urbanization, railroad travel, imperialism, territorial expansionism, and the frictions sparked by Darwinism and democratic reform. However, descriptors that may not come as readily to mind are "the Information Age" and "the Age of Communication"; nor would we likely associate this era with something called the "highway of thought," which emanated from the electric telegraph—an invention that, in retrospect, could be renamed "the Victorian Internet." Yet the rise of the electric telegraph in the mid-1800s constituted the greatest revolution in communication since the invention of the printing press in the fifteenth century and until the launch of the World Wide Web at the end of the twentieth century.

Historically, messages could be conveyed only as fast as a person could travel from one location to another. However, by the end of the eighteenth century, the Chappe brothers had constructed a rudimentary optical telegraph on a hilltop tower in France. Using the large, jointed arms attached to the roof, operators could form various configurations to communicate a message to a similar tower farther away, which would then relay the message to a third tower, and so on, in a long-distance chain. Nevertheless, even with telescopes, visibility severely hampered the efficacy of this semaphore system. After countless attempts by scientists to use electricity to transmit messages, innovators in both Britain and America, including the painter and polymath Samuel F.B. Morse, worked at harnessing electromagnetic forces to send communications via cable. But while in Britain the technology was initially reserved for railway signaling, in the U.S. it culminated in the transmission of a message along a 40-mile telegraphic wire from Washington, D.C., to Baltimore in 1844. Using a binary code of short and long electrical impulses, or "dots and dashes," Morse dispatched the words: "What hath God wrought?"

Within 20 years, telegraph cables crisscrossed the continental U.S. and much of Europe. The telegraph office became ubiquitous, and telegrams—often bouncing from one office to another like emails tossed from server to server—reached ever more remote destinations. Following some spectacular failures, a durable cable was stretched across the floor of the Atlantic Ocean, enabling Europe and the U.S. to exchange messages within minutes. In the 1870s, the British Empire connected London with outposts in India and Australia. While individual telegram speed remained relatively constant, an endless deluge of information began to pour through the wires. There was, as Tom Standage observes, "an irreversible acceleration in the pace of business life," reflecting how "telegraphy and commerce thrived in a virtuous circle." It was as though "rapid long-distance communication had effectively obliterated time and space," begetting the phenomenon known as "globalization."

A new type of skilled worker, the telegrapher, was born. He or she belonged to a vast, online community, whose semi-anonymous members shared a unique intermediary role as well as a language of dots and dashes—vaguely prefiguring the exchange of bits and bytes along modern computer networks. Like today's online communities, these telegraphers fostered their own subculture: jokes and anecdotes flew over the wires; some operators invented systems to play games; and occasional romances blossomed. Regrettably, hackers, scammers, and shady entrepreneurs also frequented the byways of this early internet.

Predictably, newer forms of the original telegraphic technology—first the telephone and, much later, the fax machine—eventually encroached. The last telegram departed from India in 2013, but the twenty-first century has arguably seen a revival of this communication form in the text message. As Standage claims, texting has not only "reincarnate[d] a defunct nineteenth-century technology" but reinforced "the democratization of telecommunications" inaugurated by the miraculous Victorian Internet.

*The Victorian Internet* ©UWorld

Annotations for this passage can be found in the CARS Passage Booklet.

### 1e. Identifying Passage Perspectives Question 2

In the last paragraph, the author compares telegrams with text messages. Based on this comparison, the author considers the telegram to be a communication mode that was:

A. elitist.
B. commonplace.
C. overrated.
D. unique.

See next page for the *strategy-based explanation* of this question.

> **1e. Identifying Passage Perspectives Question 2**
> **Strategy-based Explanation**
>
> In the last paragraph, the author compares telegrams with text messages. Based on this comparison, the author considers the telegram to be a communication mode that was:
>
> A. elitist.
> B. commonplace.
> C. overrated.
> D. unique.

In looking at this question, you may notice that it provides you with a lot of information: it tells you exactly where to look in the passage for the relevant text and asks you to focus on the comparison presented there. Therefore, we simply need to review the details of this comparison to determine what they reveal about the author's view of the telegram.

## Applying the Method

*Passage Excerpt*

**[P4]** Predictably, newer forms of the original telegraphic technology—first the telephone and, much later, the fax machine—eventually encroached. The last telegram departed from India in 2013, but the twenty-first century has **arguably** seen a revival of this communication form in the text message. **As Standage claims,** texting has not only "reincarnate[d] a defunct nineteenth-century technology" but reinforced "the democratization of telecommunications" inaugurated by the miraculous Victorian Internet.

**Step 1: Take Note of Viewpoint Indicators**

*Viewpoint Indicator*

The term "**arguably**" signals a view that the author finds reasonable; in this case, that text messages are like a modern version of telegrams.

*Viewpoint Indicator*

With the words "**as Standage claims**," the author indicates agreement with the view being cited: that both texting and telegrams led to a "democratization" or more widespread accessibility of long-distance communication.

## Step 2: Identify the Source's Belief or Attitude

Based on the viewpoint indicators "**arguably**" and "**as Standage claims**," we can conclude that the author agrees with the following claims:

- Text messages are a modern form of the telegram.

- Both telegrams and text messaging made long-distance communication more widespread and easily accessible.

In reviewing the answer choices, only **Choice B** provides a description that could sum up this view of the telegram as widespread and easily accessible. As with text messaging today, sending or receiving a telegram was a *commonplace* occurrence in the Victorian era. Accordingly, **Choice B** represents the author's view and is the correct answer.

For an alternative method of explanation based more specifically on passage evidence, you can view this question in the UWorld Qbank (sold separately).

# Passage J: The Inkblots

For almost 100 years now, the psychological evaluation known as the Rorschach Inkblot Test has engendered much controversy, including skepticism about its value, questions about its scoring, and, especially, criticism of its interpretive methods as too subjective. Thus, the Rorschach test, which emerged from the same early twentieth-century zeitgeist that produced Einstein's physics, Freudian psychoanalysis, and abstract art, seems one of modernity's most misbegotten children. Destined never to be completely accepted or discredited, the test remains a perennial outlier in its field. Nevertheless, the inkblots' mystery and aesthetic appeal have caused them to be indelibly printed on our cultural fabric.

The now iconic inkblots were introduced to the world by Swiss psychiatrist Hermann Rorschach in his 1921 book *Psychodiagnostics*. As both director of the Herisau Asylum in Switzerland and an amateur artist, Rorschach was uniquely positioned to wed the new practice of psychoanalysis to the budding phenomenon of abstract art. For instance, reading Freud's work on dream symbolism prompted him to recall his childhood passion for a game based on inkblot art called *Klecksographie*. He was also cognizant that in a recently published dissertation, his colleague Szymon Hens had used inkblots to try to probe the imagination of research subjects; moreover, a few years earlier, the French psychologist and father of intelligence testing Alfred Binet had used them to measure creativity.

Motivated by these developments, the Herisau director decided to revisit that childhood pastime that had awakened his curiosity about how visual information is processed. In particular, he wondered why different people saw different things in the same image. Traditionally, psychoanalysts had relied on language for insights; however, as biographer Damion Searls reports, Rorschach's theories would exemplify the principle that "who we are is a matter less of what we say than of what we see." Indeed, through a process of perception termed pareidolia, the mind projects meaning onto images, detecting in them familiar objects or shapes. Consequently, what a person sees in an image reveals more about that person than about the image itself.

Rorschach experimented with countless inkblots, eventually selecting ten—five black on white, two also featuring some red, and three pastel-colored—to use with research subjects. For these perfectly symmetrical images—each of which he was said to have "meticulously designed to be as ambiguous and 'conflicted' as possible"—the primary question was always "What do *you* see?" Rorschach was especially careful to note how much attention individuals paid to various components of each inkblot (such as form, color, and a sense of movement) and whether they concentrated on details or the whole image. Having observed that his patients with schizophrenia gave distinctly different responses from the control group, Rorschach envisioned his experiment as a diagnostic tool for the disease. Nevertheless, he resisted the notion that its results could be used to assess personality. In fact, until his untimely death from a ruptured appendix in 1922, Rorschach referred to his project as an "interpretive form experiment" rather than a test. Ironically, however, by the 1960s, the Rorschach Inkblot Test was known chiefly as a personality assessment and had become the most frequently administered projective personality test in the US.

Rorschach's test has survived nearly incessant scrutiny, including a 2013 comprehensive study of all Rorschach test data and repeated revisions to its scoring, yet doubts about its validity and reliability persist. Much like the inkblots themselves—which tantalize us with the possibility of divulging the secrets of who we are and how we see the world—the test has (for better or worse) defied attempts to fix its meaning. Thus, what has been called "the twentieth century's most visionary synthesis of art and science" stands tempered by harsh criticism.

*The Inkblots* ©UWorld

Annotations for this passage can be found in the CARS Passage Booklet.

> **1e. Identifying Passage Perspectives Question 3**
>
> Given the information in the passage, the author most likely believes that the way we process visual information:
>
> A. had been an important factor in psychology long before Rorschach.
> B. was initially overemphasized by Rorschach.
> C. is an integral element of a person's psychological makeup.
> D. interferes with the psychologist's ability to gain insight into an individual's personality.

See next page for the *strategy-based explanation* of this question.

## 1e. Identifying Passage Perspectives Question 3
## Strategy-based Explanation

Given the information in the passage, the author most likely believes that the way we process visual information:

- A. had been an important factor in psychology long before Rorschach.
- B. was initially overemphasized by Rorschach.
- C. is an integral element of a person's psychological makeup.
- D. interferes with the psychologist's ability to gain insight into an individual's personality.

This question asks you to discern the author's view on a specific topic presented in the passage, how we process visual information. You may recall that in discussing the evolution of the Rorschach test, Paragraph 3 brings up theories of visual processing. Hence, that section of the passage is where we will likely find the author's beliefs about such theories.

## Applying the Method

*Passage Excerpt*

**[P3]** Motivated by these developments, the Herisau director decided to revisit that childhood pastime that had awakened his curiosity about how visual information is processed. In particular, he wondered why different people saw different things in the same image. Traditionally, psychoanalysts had relied on language for insights; however, as biographer Damion Searls reports, Rorschach's theories would exemplify the principle that "who we are is a matter less of what we say than of what we see." **Indeed**, through a process of perception termed pareidolia, the mind projects meaning onto images, detecting in them familiar objects or shapes. **Consequently**, what a person sees in an image reveals more about that person than about the image itself.

### Step 1: Take Note of Viewpoint Indicators

**Viewpoint Indicator**

The author's use of the word "**indeed**" signals two things: 1) they agree with what has just been stated; and 2) they are going to elaborate on it. Accordingly, we see that the author supports the psychological principle described, then relates that principle to a specific phenomenon (pareidolia).

**Viewpoint Indicator**

By using the word "**Consequently**," the author signals that they are going to draw a conclusion about what has just been said. Accordingly, after linking Rorschach's theories with the concept of pareidolia, they conclude that when we look at an image, what we see is based more on ourselves and our own qualities than on the image itself.

### Step 2: Identify the Source's Belief or Attitude

Based on what the viewpoint indicators "**Indeed**" and "**Consequently**" have helped us to discern, we can identify the author's beliefs as follows:

1. Given how we process visual information, what we see is more important than what we say.

2. Through pareidolia, we tend to project our own traits onto an image and see what is familiar to us. Thus, what we see reveals more about who we are than it does about the image itself.

Accordingly, the author believes that the way we process visual information reveals something about us and is intertwined with our psychology. Looking at the answer choices, we find that **Choice C**, *is an integral element of a person's psychological makeup*, most closely corresponds with this view. Therefore, **Choice C** is the correct answer.

For an alternative method of explanation based more specifically on passage evidence, you can view this question in the UWorld Qbank (sold separately).

## Passage G: Food Costs and Disease

Because frequent consumption of unhealthy foods is strongly linked with cardiometabolic diseases, one way for governments to combat those afflictions may be to modify the eating habits of the general public. Applying economic incentives or disincentives to various types of foods could potentially alter people's diets, leading to more positive health outcomes.

Utilizing national data from 2012 regarding food consumption, health, and economic status, Peñalvo et al. concluded that such price adjustments would help to prevent deaths related to cardiometabolic diseases. According to their analysis, increasing the prices of unhealthy foods such as processed meats and sugary sodas by 10%, while reducing the prices of healthy foods such as fruit and vegetables by 10%, would prevent an estimated 3.4% of yearly deaths in the U.S. Changing prices by 30% would have an even stronger effect, preventing an estimated 9.2% of yearly deaths. This data comports with that found in other countries, such as "previous modeling studies in South Africa and India, where a 20% SSB [sugar-sweetened beverage] tax was estimated to reduce diabetes prevalence by 4% over 20 years." The effects of price adjustments would be most pronounced on persons of lower socioeconomic status, as the researchers "found an overall 18.2% higher price-responsiveness for low versus high SES [socioeconomic status] groups."

This differential effect based on socioeconomic status contributes to concerns about such interventions, however. In *Harvard Public Health Review*, Kates and Hayward ask: "Well intentioned though they may be, at what point do these taxes overstep government influence on an individual's right to autonomy in decision-making? On whom does the increased financial burden of this taxation fall?" They note that taxes on sugar-sweetened beverages, for instance, "are likely to have a greater impact on low-income individuals…because individuals in those settings are more likely to be beholden to cost when making decisions about food."

However, "well intentioned though they may be," the worries that Kates and Hayward express are to some extent misguided. In particular, the idea that taxing unhealthy foods would burden those least able to afford it misses the point. Although the increased taxes would affect anyone who continued to purchase the items despite the higher prices, the goal of raising prices on unhealthy foods is precisely to dissuade people from buying them. As Kates and Hayward themselves remark, "Those in low-income environments may also be the largest consumers of obesogenic foods and therefore most likely to benefit from such a lifestyle change indirectly posed by SSB taxes." As the goal of the taxes is to promote those lifestyle changes, the financial burden objection is a non-starter.

Given this recognition, the question regarding autonomy constitutes a more substantial issue. Nevertheless, that concern also rests on a dubious assumption, as people's autonomy is not necessarily respected in the current situation either. The fact that those of lower socioeconomic status are more likely to have poorer diets suggests that such persons' food choices are the result of financial constraint, not fully autonomous, rational deliberation. Hence, by making healthy foods more affordable relative to unhealthy ones, government intervention might actually facilitate autonomous choices rather than hindering them.

On the other hand, suppose that the disproportionate consumption of unhealthy foods—and associated higher incidence of disease—among certain groups is not the result of financial hardship but rather the result of those persons' perceived self-interest. If so, that would suggest that members of these groups are being encouraged to persist in harmful dietary habits for the sake of corporate profits. In that case, violating autonomy for the sake of health may be permissible, as that practice would be morally preferable to the present system of corporate exploitation.

*Food Costs and Disease* ©UWorld

Annotations for this passage can be found in the CARS Passage Booklet.

## 1e. Identifying Passage Perspectives Question 4

According to the author, a reason it might be legitimate for government interventions to violate some citizens' autonomy would be if implementing those interventions:

- A. is supported by the available research data.
- B. is supported by a majority of all citizens.
- C. is aimed at citizens disproportionately affected by an issue.
- D. is less problematic than not implementing them.

See next page for the *strategy-based explanation* of this question.

Chapter 5: Skill 1 – Foundations of Comprehension

> ### 1e. Identifying Passage Perspectives Question 4
> ### Strategy-based Explanation
>
> According to the author, a reason it might be legitimate for government interventions to violate some citizens' autonomy would be if implementing those interventions:
>
> A. is supported by the available research data.
> B. is supported by a majority of all citizens.
> C. is aimed at citizens disproportionately affected by an issue.
> D. is less problematic than not implementing them.

If we remember how the passage is structured, we can quickly locate the author's views about the question topic. The passage initially discusses how government interventions on food prices can improve public health. Then, an objection is raised: such interventions might violate consumers' autonomy. Ultimately, the author responds to this objection in Paragraphs 5 and 6, revealing their views about when violating autonomy may be legitimate.

## Applying the Method

*Passage Excerpt*

**[P5]** Given this recognition, **the question regarding autonomy constitutes a more substantial issue**. Nevertheless, that concern also rests on **a dubious assumption**, as people's autonomy is not necessarily respected in the current situation either. The fact that those of lower socioeconomic status are more likely to have poorer diets **suggests that such persons' food choices are the result of financial constraint, not fully autonomous, rational deliberation.**

**[P6]** On the other hand, suppose that the disproportionate consumption of unhealthy foods—and associated higher incidence of disease—among certain groups is not the result of financial hardship but rather the result of those persons' perceived self-interest. **If so, that would suggest that members of these groups are being encouraged to persist in harmful dietary habits for the sake of corporate profits. In that case, violating autonomy for the sake of health may be permissible, as that practice would be morally preferable to the present system of corporate exploitation.**

### Step 1: Take Note of Viewpoint Indicators

**Viewpoint Indicator**

By describing the potential threat to consumers' autonomy as "**a more substantial issue**" (than concern about a financial burden), the author implies that they see this issue as important.

**Viewpoint Indicator**

However, the author doubts (it is "**a dubious assumption**") that many consumers have autonomy to begin with, since **their poor food choices are likely driven by finances.**

**Viewpoint Indicator**

Finally, the author argues that consumers may be **tricked by corporations into making unhealthy food choices**. In that case, the author concludes:

It is **morally preferable** for governments to violate autonomy to improve people's health than to allow corporations to exploit people for profits.

## Step 2: Identify the Source's Belief or Attitude

Using the viewpoint indicators we recognized, we could identify the author's beliefs about citizens' autonomy. Essentially, the author makes the following argument:

1. Autonomy is important, but it is doubtful that many citizens have it in the first place. Even if they do, they are likely being exploited for profit by corporations.

2. Exploiting people for profit is morally worse than violating people's autonomy to improve their health. Therefore, violating citizens' autonomy is the better alternative.

Accordingly, we can see that **Choice D**, *is less problematic than not implementing them*, reflects the author's position. The author makes a comparison: violating autonomy and exploitation are both harmful, but one action improves the citizens' health while the other uses the citizens for financial gain. In other words, the author argues that *implementing government interventions would be less problematic than not implementing them*, making **Choice D** the correct answer.

For an alternative method of explanation based more specifically on passage evidence, you can view this question in the UWorld Qbank (sold separately).

Lesson 5.6

# CMP Subskill 1f. Further Implications of Passage Claims

| Skill 1: Foundations of Comprehension ||
|---|---|
| **Subskill** | **Student Objective** |
| 1a. Main Idea or Purpose | Identify the central theme or purpose of the passage |
| 1b. Meaning of Term | Determine the meaning of a term as used in the passage |
| 1c. Direct Passage Claims | Identify claims made directly in the passage |
| 1d. Implicit Claims or Assumptions | Identify implicit claims or unstated assumptions in the passage |
| 1e. Identifying Passage Perspectives | Identify perspectives or positions presented in the passage |
| **1f. Further Implications of Passage Claims** | **Recognize further implications of claims stated in the passage** |

The final subskill for Foundations of Comprehension questions is **1f. Further Implications of Passage Claims**. To answer these types of questions, you must recognize conclusions that can be drawn from existing passage information. These answers are not themselves in the passage but can be logically inferred from what the passage states.

## Method for Answering 1f. Direct Passage Claims Questions

### Step 1: Verify What the Passage Directly States

As with **1c** and **1d** questions, the first step in approaching **1f** questions is to review what the passage says about the question topic. The answer to the question will go slightly beyond this information to draw a conclusion that follows from those passage claims.

### Step 2: Ask What Additional Claims Would Follow from Existing Statements

To draw this conclusion, you must consider what additional statements would be true if passage information is correct. In other words, given what you know already, what must also be the case? This process is less mysterious than it may sound; in fact, you probably engage in it often without being aware. For instance, whenever you think something like: "If that's true, it must also mean that..." you are inferring such additional claims.

### Step 3: Compare Answer Choices with the Additional Claims

The final step is to compare the answer choices with those claims. The correct answer will be a logical extension of passage information; thus, while this answer is not included in the passage, it should seem as if it could have been.

---

**Subskill Connection: 1f / 2f / 3f**

As noted in Lesson 4.4, there are three different subskills that involve drawing new conclusions, each corresponding to one of the overarching CARS skills: **1f. Further Implications of Passage Claims** (Foundations of Comprehension); **2f. Drawing Additional Inferences** (Reasoning Within the Text); and **3f. Additional Conclusions From New Information** (Reasoning Beyond the Text).

These three types of questions are similar, but vary in their complexity. For example, **1f. Further Implications of Passage Claims** questions may require fewer steps in reasoning than are needed to answer **2f. Drawing Additional Inferences** questions. **3f. Additional Conclusions From New Information** questions go a step further by introducing outside information that must also be considered. These subskills are further compared in Lessons 6.6 and 7.6.

## Subskill 1f. Further Implications of Passage Claims
### (3 Practice Questions)

We now turn to three examples from the UWorld Qbank. In each case, the passage and question are first given without commentary, allowing you to practice applying the method yourself. Then, the passage and question are presented again, this time with annotations of the passage (see the CARS Passage Booklet) and a step-by-step explanation of the question using the described method.

## Passage K: Lengthening the School Day

There may be reasons to reject the idea of lengthening the school day. None of them, however, are *good* reasons. Rather, the supposed demerits of such a proposal fall easily in the face of its numerous financial and social benefits for families.

The greatest of these benefits lies in reducing the need for childcare. It is a curious fact of American life that the adult's work schedule and the child's school schedule are misaligned. Children rise with the sun to head to classes, only to be sent home again hours before parents return from their jobs. In a society where, more often than not, both parents work, this discordance creates the need for an expensive arrangement to fill the gap in families' routines. For instance, studies show that in 2016, childcare costs accounted for 9.5 to 17.5 percent of median family income, depending on the state. Today, 40 percent of families nationwide spend over 15 percent of their income on childcare. Transportation to and from care sites only adds to that expense.

An additional advantage of an extended school day would be to allow for greater diversity and depth in curricula. Schools across the country have increasingly cut instruction in arts, music, and physical education (as well as recess) in order to meet objectives in reading and math. While this unfortunate state of affairs can be partially blamed on overzealous attention to standardized tests, it points to the larger deleterious trend of narrowing students' instruction. With a longer school day, such eliminated subjects can be restored, enriching students with a more well-rounded education.

To this proposal, however, critics may object that the added time would impose strain on educators. Can we truly ask schoolteachers—already among the most overworked individuals in society—to endure even more hours in the classroom? The answer is that a lengthened school day need not distress teachers nor add to their already cumbersome workload. By providing for additional areas of study in the arts and humanities, the extension would give schools cause to hire new, perhaps specialized, faculty to offer these courses. Moreover, the time could also be allocated to sports, academic clubs, and other extracurricular activities.

However, this point speaks to another objection, namely, the cost of adjusting the school day. Whether through paying current teachers more or hiring new ones, implementing such a proposal would entail a significant financial expenditure. There are at least two responses to this line of thinking. First is that this increase in the cost of schooling would be offset and likely surpassed by the aforementioned savings in childcare. Thus, while it is true that schools would require greater funding (likely necessitating higher property taxes), parents would ultimately pay the same or less overall, with greater educational opportunities for their children and fewer transportational burdens. Second is that schools should be better funded regardless. Recently, some schools—especially those in rural areas—have even reduced school weeks to only four days as a cost-saving measure. It is beyond dispute that schools across the board both need and deserve a radically increased investment from citizens. Lengthening the school day is simply one manifestation of how such funding should be utilized.

With this one change, states can coordinate the lives of parents and children, reduce the need for costly childcare, and expand curricular offerings. These worthy and desirable aims provide a clear justification for extending the school day.

*Lengthening the School Day* ©UWorld

Annotations for this passage can be found in the CARS Passage Booklet.

### 1f. Further Implications of Passage Claims Question 1

Based on the information presented in the passage, one could most reasonably conclude that a significant portion of the funding a school receives is based on the:

A. amount of time spent in the classroom.
B. local need for childcare.
C. value of homes and other buildings.
D. breadth of its curriculum.

See next page for the *strategy-based explanation* of this question.

Chapter 5: Skill 1 – Foundations of Comprehension

> ### 1f. Further Implications of Passage Claims Question 1
> ### Strategy-based Explanation
>
> Based on the information presented in the passage, one could most reasonably conclude that a significant portion of the funding a school receives is based on the:
>
> A. amount of time spent in the classroom.
> B. local need for childcare.
> C. value of homes and other buildings.
> D. breadth of its curriculum.

If you were to encounter this question on the exam, the phrase "one could most reasonably conclude" would be a clue to help you identify the type of question it is. This wording shows you are being asked to draw a conclusion about how schools are funded that would be justified by passage information but that probably isn't stated in the passage.

You may recall that the passage is structured to first describe the benefits of lengthening the school day, then address possible objections to doing so. One of those objections concerns the potential costs involved. So, if you weren't sure where school funding is mentioned, you could guess that the discussion about costs would be a good place to look for that information.

## Applying the Method

*Passage Excerpt*

**[P5]** However, this point speaks to another objection, namely, the cost of adjusting the school day. Whether through paying current teachers more or hiring new ones, implementing such a proposal would entail a significant financial expenditure. There are at least two responses to this line of thinking. First is that this increase in the cost of schooling would be offset and likely surpassed by the aforementioned savings in childcare. Thus, while it is true that schools **would require greater funding (likely necessitating higher property taxes)**, parents would ultimately pay the same or less overall, with greater educational opportunities for their children and fewer transportational burdens. Second is that **schools should be better funded regardless.** Recently, some schools—especially those in rural areas—have even reduced school weeks to only four days as a cost-saving measure. It is beyond dispute that schools across the board both need and deserve a radically increased investment from citizens. **Lengthening the school day is**

### Step 1: Verify What the Passage Directly States

The author discusses the cost of lengthening the school day in Paragraph 5. Within that discussion, they make three specific claims about school funding:

1. **If schools receive more funding, then property taxes will probably have to increase.**

2. **Schools should receive more funding regardless of whether the school day is lengthened.**

3. **Lengthening the school day is only one potential benefit of increased school funding.**

However, only the first of these claims gives us any additional information about school funding: namely, that it has something to do with **property taxes**. The second claim says that schools **should receive more funding**, and the third talks about the **benefits of more funding**,

**simply one manifestation of how such funding should be utilized.** | but neither of them say anything about where that funding comes from.

### Step 2: Ask What Additional Claims Would Follow from Existing Statements

We have now identified the claim most relevant to our question:

1. **If schools receive more funding, then property taxes will probably have to increase.**

If that statement is true, then we can draw the following conclusion:

Therefore, **property taxes must be the source for some or all of the funding a school receives**.

Accordingly, the answer to the question should be a claim that is similar to this conclusion.

### Step 3: Compare Answer Choices with the Additional Claims

Given the relevant passage information, we were able to determine that **property taxes must be the source for some or all of the funding a school receives**.

Turning to the answer choices, we see that **Choice C**, *value of homes and other buildings*, similarly refers to types of properties and how much they are worth. Accordingly, this phrase represents a more specific version of the conclusion we drew in Step 2: **property taxes *on homes and other buildings* must be the source for some or all of the funding a school receives**.

Therefore, **Choice C** represents a further implication of passage claims and is the correct answer.

For an alternative method of explanation based more specifically on passage evidence, you can view this question in the UWorld Qbank (sold separately).

## Passage F: The Victorian Internet

Lasting roughly from 1820 to 1914, the Victorian Era is often defined by its many distinctive sociological conditions, including industrialization, urbanization, railroad travel, imperialism, territorial expansionism, and the frictions sparked by Darwinism and democratic reform. However, descriptors that may not come as readily to mind are "the Information Age" and "the Age of Communication"; nor would we likely associate this era with something called the "highway of thought," which emanated from the electric telegraph—an invention that, in retrospect, could be renamed "the Victorian Internet." Yet the rise of the electric telegraph in the mid-1800s constituted the greatest revolution in communication since the invention of the printing press in the fifteenth century and until the launch of the World Wide Web at the end of the twentieth century.

Historically, messages could be conveyed only as fast as a person could travel from one location to another. However, by the end of the eighteenth century, the Chappe brothers had constructed a rudimentary optical telegraph on a hilltop tower in France. Using the large, jointed arms attached to the roof, operators could form various configurations to communicate a message to a similar tower farther away, which would then relay the message to a third tower, and so on, in a long-distance chain. Nevertheless, even with telescopes, visibility severely hampered the efficacy of this semaphore system. After countless attempts by scientists to use electricity to transmit messages, innovators in both Britain and America, including the painter and polymath Samuel F.B. Morse, worked at harnessing electromagnetic forces to send communications via cable. But while in Britain the technology was initially reserved for railway signaling, in the U.S. it culminated in the transmission of a message along a 40-mile telegraphic wire from Washington, D.C., to Baltimore in 1844. Using a binary code of short and long electrical impulses, or "dots and dashes," Morse dispatched the words: "What hath God wrought?"

Within 20 years, telegraph cables crisscrossed the continental U.S. and much of Europe. The telegraph office became ubiquitous, and telegrams—often bouncing from one office to another like emails tossed from server to server—reached ever more remote destinations. Following some spectacular failures, a durable cable was stretched across the floor of the Atlantic Ocean, enabling Europe and the U.S. to exchange messages within minutes. In the 1870s, the British Empire connected London with outposts in India and Australia. While individual telegram speed remained relatively constant, an endless deluge of information began to pour through the wires. There was, as Tom Standage observes, "an irreversible acceleration in the pace of business life," reflecting how "telegraphy and commerce thrived in a virtuous circle." It was as though "rapid long-distance communication had effectively obliterated time and space," begetting the phenomenon known as "globalization."

A new type of skilled worker, the telegrapher, was born. He or she belonged to a vast, online community, whose semi-anonymous members shared a unique intermediary role as well as a language of dots and dashes—vaguely prefiguring the exchange of bits and bytes along modern computer networks. Like today's online communities, these telegraphers fostered their own subculture: jokes and anecdotes flew over the wires; some operators invented systems to play games; and occasional romances blossomed. Regrettably, hackers, scammers, and shady entrepreneurs also frequented the byways of this early internet.

Predictably, newer forms of the original telegraphic technology—first the telephone and, much later, the fax machine—eventually encroached. The last telegram departed from India in 2013, but the twenty-first century has arguably seen a revival of this communication form in the text message. As Standage claims, texting has not only "reincarnate[d] a defunct nineteenth-century technology" but reinforced "the democratization of telecommunications" inaugurated by the miraculous Victorian Internet.

*The Victorian Internet* ©UWorld

Annotations for this passage can be found in the CARS Passage Booklet.

### 1f. Further Implications of Passage Claims Question 2

Based on the passage, the comparison between Victorian and modern online subcultures suggests:

A. that even when two cultures are distinct, their online subcultures tend to be similar.
B. that the development of a subculture is the most significant effect of online communication.
C. that an online subculture is defined solely by a shared language.
D. that the transmission of messages cannot be trusted due to online criminal activity.

See next page for the *strategy-based explanation* of this question.

> ## 1f. Further Implications of Passage Claims Question 2
> ### Strategy-based Explanation
>
> Based on the passage, the comparison between Victorian and modern online subcultures suggests:
>
> A. that even when two cultures are distinct, their online subcultures tend to be similar.
> B. that the development of a subculture is the most significant effect of online communication.
> C. that an online subculture is defined solely by a shared language.
> D. that the transmission of messages cannot be trusted due to online criminal activity.

In approaching this question, you might recall the passage's title, "The Victorian Internet," which implies an overall comparison between the nineteenth-century telegraph system and the modern internet. This question asks you to consider a specific aspect of that comparison and what it suggests.

## Applying the Method

*Passage Excerpt*

**Step 1: Verify What the Passage Directly States**

[P4] A new type of skilled worker, the telegrapher, was born. He or she belonged to a vast, online community, whose semi-anonymous members shared a unique intermediary role as well as a language of dots and dashes—vaguely prefiguring the exchange of bits and bytes along modern computer networks. **Like today's online communities, these telegraphers fostered their own subculture: jokes and anecdotes flew over the wires; some operators invented systems to play games; and occasional romances blossomed. Regrettably, hackers, scammers, and shady entrepreneurs also frequented the byways of this early internet.**

Paragraph 4 compares the **subculture of Victorian telegraphers** with that of **today's online communities**. Like current internet users, these telegraphers engaged in **jokes, games, and romances**, while also encountering **hackers and online scams**.

### Step 2: Ask What Additional Claims Would Follow from Existing Statements

The description in Paragraph 4 lists several features of telegraph subculture that resemble how people experience the modern internet. Accordingly, we can conclude that despite the differences in time period and level of technology, **the Victorian telegraph system and the modern-day internet generated similar subcultures**.

### Step 3: Compare Answer Choices with the Additional Claims

Based on the comparison between users of the telegraph and the modern internet, we concluded that despite the differences in time period and level of technology, **the Victorian telegraph system and the modern-day internet generated similar subcultures**.

Looking at the answer choices, **Choice A** conveys a more general version of the same notion: *that even when two cultures are distinct, their online subcultures tend to be similar*. Accordingly, this statement represents a logical extension of passage information and is the correct answer.

For an alternative method of explanation based more specifically on passage evidence, you can view this question in the UWorld Qbank (sold separately).

## Passage H: For Whom the Bell Toils

In nineteenth-century America, most people dismissed the notion that someone might assassinate the president. The presumption was based not only on ethics but practicality: a president's term is inherently limited, and an unpopular one could be voted out of office. Therefore, it was reasoned, there would be no need to consider removal through violence. This belief persisted even after the shocking murder of Abraham Lincoln in 1865, which was viewed as an aberration. Thus it was that on July 2, 1881, Charles Guiteau could simply walk up to President James A. Garfield and shoot him in broad daylight. As Richard Menke portrays events, "Guiteau was in fact a madman who had come to identify with a disgruntled wing of the Republican Party after his deranged fantasies of winning a post from the new administration had come to nothing." Believing that God had told him to kill the president, Guiteau thought this act would garner fame for his religious ideas and thereby help to usher in the Apocalypse.

In an interesting parallel, Garfield had felt a sense of divine purpose for his own life after surviving a near-drowning as a young man. Unlike Guiteau's fanatical ravings, however, Garfield's vision worked to the betterment of himself and the world. Candice Millard describes his ascent from extreme poverty to incredible excellence in college, where "by his second year…they made him a professor of literature, mathematics, and ancient languages." Garfield would go on to join the Union Army, where he attained the rank of major general and argued that black soldiers should receive the same pay as their white compatriots. While serving in the war he was nominated for the House of Representatives but accepted the seat only after President Lincoln declared that the country had more need of him as a congressman than as a general. The reluctant politician would later himself become president under similar circumstances, after multiple factions of a deadlocked Republican convention unexpectedly nominated him instead of their original candidates in 1880. An honest man who opposed corruption within the party, Garfield strove both to heal the fractures of the Civil War and to uphold the aims for which it was fought, until "the equal sunlight of liberty shall shine upon every man, black or white, in the Union."

Although Guiteau's bullet would ultimately dim this light for Garfield, the president actually survived the initial attack and for a time appeared headed for recovery. Tragically, however, the hubris shown by his main physician, Dr. Willard Bliss, would lead instead to weeks of prolonged suffering. None of the doctors who examined Garfield were able to locate the bullet, and its lingering presence—along with the unwashed hands of the doctors who probed for it—led to an infection. As the president's condition worsened, inventor Alexander Graham Bell attempted to adapt his patented telephone technology to locate foreign metal in the human body. Inspired by speculation that the bullet's electromagnetic properties might be detectable, Bell used his newly developed "Induction Balance" device to listen for the sounds of electrical interference he hoped would isolate the site of the bullet.

Unfortunately, Bell's searches were unsuccessful. Like Garfield's doctors, he had been looking for the bullet in the wrong area. Menke asserts that Bell's efforts "would probably have fallen short" regardless. However, other historians suggest that Dr. Bliss, unwilling to consider challenges to his original assessment, prevented Bell from more thoroughly searching the president's body. Certainly, Bliss ignored the advice and protestations of other physicians, even as Garfield continued to decline. With death imminent, Garfield asked to be taken to his seaside cottage, where he died on the 19th of September.

*For Whom the Bell Toils* ©UWorld

Annotations for this passage can be found in the CARS Passage Booklet.

### 1f. Further Implications of Passage Claims Question 3

In his ascent to the presidency, Garfield was probably LEAST motivated by which of the following?

A. Duty
B. Justice
C. Service
D. Ambition

See next page for the *strategy-based explanation* of this question.

> ### 1f. Further Implications of Passage Claims Question 3
> ### Strategy-based Explanation
>
> In his ascent to the presidency, Garfield was probably LEAST motivated by which of the following?
>
> A. Duty
> B. Justice
> C. Service
> D. Ambition

While some "least" questions require extensive passage review, this one asks about Garfield's ascent to the presidency, which is discussed in only one section. Specifically, Paragraph 2 describes the circumstances under which Garfield became president, so we can focus on that information to determine what it suggests about his motivations.

## Applying the Method

*Passage Excerpt*

[P2] Garfield would go on to join the Union Army, where he attained the rank of major general and argued that black soldiers should receive the same pay as their white compatriots. While serving in the war **he was nominated** for the House of Representatives but **accepted the seat only after President Lincoln declared that the country had more need of him as a congressman than as a general.** The **reluctant politician** would later himself **become president under similar circumstances**, after multiple factions of a deadlocked Republican convention unexpectedly nominated him instead of their original candidates in 1880.

**Step 1: Verify What the Passage Directly States**

In the author's description of Garfield's political career, we find two related observations:

1. Garfield joined Congress only after President Lincoln said the country needed him there.

2. Garfield became president only after other Republicans said they needed him to run.

Therefore, the passage tells us that Garfield's ascent to the presidency was like his ascent to Congress: **He didn't set out to hold office, but served at the request of other people.**

### Step 2: Ask What Additional Claims Would Follow from Existing Statements

The passage states that **Garfield had not intended to seek the presidency but agreed to serve that role when asked**.

Based on that claim, we can draw the following conclusion about Garfield's motivations:

> Therefore, **in seeking the presidency, Garfield was motivated more by serving others than by his own desires**.

Accordingly, the answer to the question should reflect this conclusion.

### Step 3: Compare Answer Choices with the Additional Claims

Based on the passage claims about Garfield's political career, we concluded that **in seeking the presidency, Garfield was motivated more by serving others than by his own desires**. However, the question asks what probably LEAST motivated Garfield, so the correct answer should represent the opposite: being motivated by one's own desires rather than serving others.

Accordingly, the answer that stands out is **Choice D,** *ambition*, which describes being driven by one's own desire for self-advancement. By contrast, all three of the other choices relate to serving the country or other people. Therefore, **Choice D** least matches Garfield's motivations and is the correct answer.

For an alternative method of explanation based more specifically on passage evidence, you can view this question in the UWorld Qbank (sold separately).

Lesson 6.1

# RWT Subskill 2a. Logical Relationships Within Passage

The second competency or skill of the CARS section is **Reasoning Within the Text (RWT)**. These types of questions concern how various passage information is related. For example, questions may ask how different passage claims are connected, or what the reader can infer from information found in multiple parts of the passage.

Like CMP questions, RWT questions can be classified into six types that correspond to specific subskills. For each subskill, a description is given, followed by a method for approaching such questions. Finally, example questions are provided to illustrate the method and give you practice.

| Skill 2: Reasoning Within the Text ||
|---|---|
| **Subskill** | **Student Objective** |
| **2a. Logical Relationships Within Passage** | **Identify logical relationships between passage claims** |
| 2b. Function of Passage Claim | Determine what function a particular claim serves in the passage |
| 2c. Extent of Passage Evidence | Determine the extent to which the passage provides evidence for its claims |
| 2d. Connecting Claims With Evidence | Connect passage claims with supporting passage evidence |
| 2e. Determining Passage Perspectives | Determine the perspective of the author or another source in the passage |
| 2f. Drawing Additional Inferences | Draw additional inferences based on passage information |

**2a. Logical Relationships Within Passage** questions are concerned with the reasoning behind passage claims and how those claims fit together. For instance, these questions may ask about:

- **The information a particular claim depends on**

    "In the author's thought experiment, which of the following questions would be most relevant to determining whether Mark lied about the meeting?"

- **Weaknesses of an argument**

    "Which of the following statements regarding the author's argument would represent the most reasonable criticism?"

- **Whether different sections of the passage contradict each other**

    "Which of the following ideas is the author *inconsistent* about in the passage?"

- **How a statement connects to a larger theme**

    "The description of socialism as 'a pervasive bogeyman' is part of the author's claim that:"

Accordingly, you might think of these questions as asking about the links between various parts of the passage.

## Method for Answering 2a. Logical Relationships Within Passage Questions

### Step 1: Restate the Relevant Claim or Argument

The first step in answering these questions is to restate the claim or argument they ask about in simpler terms. This process will help refamiliarize you with the relevant passage information, making it easier to see the connections between passage ideas.

### Step 2: Identify the Key Factors

After restating that claim or argument, the next step is to note its components and how they relate to each other. For example, accepting a particular passage assertion might presuppose that another assertion is true; the author might use one claim to justify another; or two statements in the passage taken together might imply a new conclusion. Recognizing these connections will reveal the factors on which the claim or argument most depends.

### Step 3: Compare Answer Choices with the Key Factors

Finally, the key factors you have identified will address the subject that the question asks about. You will find that one answer choice closely reflects these factors, explaining the relevant logical relationship and thereby answering the question.

## Subskill 2a. Logical Relationships Within Passage
### (4 Practice Questions)

We now look at four examples from the UWorld Qbank. In each case, the passage and question are first given without commentary, allowing you to practice applying the method yourself. Then, the passage and question are presented again, this time with annotations of the passage (see the CARS Passage Booklet) and a step-by-step explanation of the question using the described method.

## Passage C: Probability and The Universe

The idea of probability is frequently misunderstood, in large part because of a conceptual confusion between objective probability and subjective probability. The failure to make this distinction leads to an erroneous conflation of genuine possibility with what is in fact merely personal ignorance of outcome. An example will clarify.

A standard die is rolled on a table, but the outcome of the roll is concealed. Should an observer be asked the chance that a particular number was rolled—five, say—the natural response is 1/6. However, this answer is incorrect. To say there is a one-in-six chance that a five was rolled implies there is an equal chance that any of the other numbers were rolled. But there is no equal chance, because the roll has already occurred. Hence, the probability that the result of the roll is a five is either 100% or 0%, and the same is true for each of the other numbers.

It might be objected that such an analysis is an issue of semantics rather than a substantive claim. For declaring the probability to be 1/6 is merely an expression that, for all we know, any number from 1 through 6 might have been rolled. But the difference between *for all we know* and what *is* remains crucially important, because it forestalls a tendency to make scientific assertions from a perspective biased by human perception….

[T]hus, it should be clear that claims about the purpose of the universe rest on shaky ground. In particular, we must be wary of inferences drawn from juxtaposing the existence of intelligent life with the genuinely improbable cosmological conditions that make such life possible. Joseph Zycinski provides an instructive account of those conditions: "If twenty billion years ago the rate of expansion were minimally smaller, the universe would have entered a stage of contraction at a time when temperatures of thousands of degrees were prevalent and life could not have appeared. If the distribution of matter were more uniform the galaxies could not have formed. If it were less uniform, instead of the presently observed stars, black holes would have formed from the collapsing matter." In short, the existence of life (let alone intelligent life) depends upon the initial conditions of the universe having conformed to an extremely narrow range of possible values.

It is tempting to go beyond Zycinski's factual point to draw deep cosmological, teleological, and even theological conclusions. But no such conclusions follow. The reason why a life-sustaining universe exists is that if it did not, there would be no one to wonder why a life-sustaining universe exists. This fact is a function not of purpose, but of pre-requisite. For instance, suppose again that a five was rolled on a die. We now observe a five on its face, not because the five was "meant to be" but because one side of the die had to land face up. Even if we imagine that the die has not six but six *billion* sides, that analysis is unaffected. Yes, it was unlikely that any given value would be rolled, but *some* value had to be rolled. That "initial condition" then set the parameters for what kind of events could possibly follow; in this case, the observation of the five.

Similarly, the existence of intelligent beings is evidence that certain physical laws obtained in the universe. However, it is not evidence that those beings or laws were necessary or intended rather than essentially random. The subjective probability of that outcome is irrelevant, because in this universe the objective probability of its occurrence is 100%. Thus, the seemingly low initial chance that such a universe would exist is not in itself indicative of a purpose to its existence.

*Probability and The Universe* ©UWorld

Annotations for this passage can be found in the CARS Passage Booklet.

> ## 2a. Logical Relationships Within Passage Question 1
>
> How could the author more clearly demonstrate the point that, in the die-rolling example, "the probability that the result of the roll is a five is either 100% or 0%"?
>
> A. Allowing multiple rolls to be made in order to give repeated results
> B. Allowing a die with more than six sides to be used for the roll
> C. Allowing observers to see the result of the die roll instead of concealing it
> D. Allowing observers to predict the probability of a five being rolled before the roll is actually made

See next page for the *strategy-based explanation* of this question.

> ## 2a. Logical Relationships Within Passage Question 1
> ### Strategy-based Explanation
>
> How could the author more clearly demonstrate the point that, in the die-rolling example, "the probability that the result of the roll is a five is either 100% or 0%"?
>
> A. Allowing multiple rolls to be made in order to give repeated results
> B. Allowing a die with more than six sides to be used for the roll
> C. Allowing observers to see the result of the die roll instead of concealing it
> D. Allowing observers to predict the probability of a five being rolled before the roll is actually made

This question asks how one of the author's points could be made clearer. Therefore, we should proceed by examining how the point is originally presented and breaking down the claims the author makes in presenting it. We can then restate those claims in a way that highlights the logical connections between them.

## Applying the Method

*Passage Excerpt*

**Step 1: Restate the Relevant Claim or Argument**

**[P2]** A standard die is rolled on a table, but **the outcome of the roll is concealed**. Should an observer be asked the chance that a particular number was rolled—five, say—**the natural response is 1/6. However, this answer is incorrect.** To say there is a one-in-six chance that a five was rolled implies there is an equal chance that any of the other numbers were rolled. But **there is no equal chance, because the roll has already occurred. Hence, the probability that the result of the roll is a five is either 100% or 0%**, and the same is true for each of the other numbers.

After reviewing the author's argument about the die-rolling example, we can break it down as follows:

1. **If you don't know the outcome of a die roll, you naturally think there is a 1/6 chance that a five came up.**

2. **But since the roll has already happened, five either came up or it didn't.**

3. **So, the chance isn't 1/6, it's either 100% or 0%.**

Now we have a more straightforward presentation of the author's argument. Statement 3 is a conclusion that the author draws from statements 1 and 2.

### Step 2: Identify the Key Factors

Now that we have restated the author's argument, we can see how its components fit together. In particular, we notice that the argument begins with a specific condition:

1. **If you don't know the outcome of a die roll, you naturally think there is a 1/6 chance that a five came up**.

Why is this condition important? Or, returning to the passage text, why does the author start by specifying that **the outcome of the roll is concealed**?

To answer that question, suppose we remove the condition. Now our scenario is: a die is rolled, you see the result, and you are asked what the chance is that a five came up. In this situation, would 1/6 seem like the correct answer? No, because you would *know* whether the five came up or not. In other words, you would see either that the "chance" a five came up was 100% or that it was 0%. Accordingly, **not knowing the outcome** of the die roll is a key factor in the author's point; that ignorance is the only reason someone would think the chance of a five was still 1/6.

### Step 3: Compare Answer Choices with the Key Factors

We have determined that **not knowing the outcome** of the die roll is the key factor causing someone to make a mistake about its probability. Therefore, removing that factor should make it easy for the author to demonstrate how the probability of the roll should actually be understood.

Looking at the answer choices, **Choice C**, *Allowing observers to see the result of the die roll instead of concealing it*, does remove that factor. If the result of the die roll is not concealed, observers would have no reason to think there was a 1/6 chance of a five; instead, they could see for themselves either that the chance was 100% or that it was 0%. In that case, the author's point about the probability of the die roll would be clearly demonstrated, making **Choice C** the correct answer.


For an alternative method of explanation based more specifically on passage evidence, you can view this question in the UWorld Qbank (sold separately).

## Passage D: American Local Motives

Locomotives were invented in England, with the first major railroad connecting Liverpool and Manchester in 1830. However, it was in America that railroads would be put to the greatest use in the nineteenth century. On May 10, 1869, the Union Pacific and Central Pacific lines met at Promontory Point, Utah, joining from opposite directions to complete a years-long project—the Transcontinental Railroad. This momentous event connected the eastern half of the United States with its western frontier and facilitated the construction of additional lines in between. As a result, journeys that had previously taken several months by horse and carriage now required less than a week's travel. By 1887 there were nearly 164,000 miles of railroad tracks in America, and by 1916 that number had swelled to over 254,000.

While the United States still has the largest railroad network in the world, it operates largely in the background of American life, and citizens no longer view trains with the sense of importance those machines once commanded. Nevertheless, the economic and industrial advantages those citizens enjoy today would not have been possible without America's history of trains; as Tom Zoellner reminds us, "Under the skin of modernity lies a skeleton of railroad tracks." Although airplanes and automobiles have now assumed greater prominence, the time has arrived for the resurgence of railroads. A revitalized and advanced railway system would confer numerous essential benefits on both the United States and the globe.

The chief obstacles to garnering support for such a project are the current dominance of the automobile and the languishing technology of existing railroads. In a sense these two obstacles are one, as American dependence on personal automobiles is partially due to the paucity of rapid public transportation. The railroads of Europe and Japan, by comparison, have vastly outpaced their American counterparts. Japan has operated high-speed rail lines continuously since 1964, and in 2007, a French train set a record of 357 miles per hour. While that speed was achieved under tightly controlled conditions, it still speaks to the great disparity in railroad development between the United States and other countries since the mid-twentieth century. British trains travel at speeds much higher than those in America, where both the trains themselves and the infrastructure to support them have simply been allowed to fall behind. In much of Europe it is common for trains to travel at close to 200 miles per hour.

To invest in a modern network of railroads would improve the United States in much the same way that the first railroads did in the nineteenth and early twentieth centuries. A high-speed passenger rail system would dramatically transform American life as travel between cities and states became quicker and more convenient, encouraging commerce, business, and tourism. Such a system would also make important strides in environmental preservation. According to a 2007 British study, "CO2 emissions from aircraft operations are...at least five times greater" than those from high-speed trains. For similar reasons, Osaka, Japan, was ranked as "the best…green transportation city in Asia" by the 2011 *Green City Index*. As Lee-in Chen Chiu notes in *The Kyoto Economic Review*, Osakans travel by railway more than twice as much as they travel by car.

It is true that developing a countrywide high-speed rail system would come with significant costs. However, that was also true of the original Transcontinental Railroad, as indeed it is with virtually any great project undertaken for the public good. We should thus move ahead with confidence that the rewards will outweigh the expenditure as citizens increasingly choose to travel by train. Both for society's gain and the crucial well-being of the planet, our path forward should proceed upon rails.

*American Local Motives* ©UWorld

Annotations for this passage can be found in the CARS Passage Booklet.

## 2a. Logical Relationships Within Passage Question 2

Which of the following passage assertions is the most essential for readers to accept in order for them to find the author's argument reasonable?

A. The airline and car industries have undue influence over American travel.
B. A high-speed railroad network would encourage economic activity.
C. The superiority of foreign railroads disadvantages America in international trade.
D. The public would readily adopt high-speed rail travel if it were available.

See next page for the *strategy-based explanation* of this question.

## 2a. Logical Relationships Within Passage Question 2
### Strategy-based Explanation

Which of the following passage assertions is the most essential for readers to accept in order for them to find the author's argument reasonable?

A. The airline and car industries have undue influence over American travel.
B. A high-speed railroad network would encourage economic activity.
C. The superiority of foreign railroads disadvantages America in international trade.
D. The public would readily adopt high-speed rail travel if it were available.

This question is somewhat oddly worded, but we can rephrase it as: "What would I most need to believe for the author's argument to be convincing?" To determine the answer, we need to review the author's argument in the passage and identify the claim on which it most depends.

## Applying the Method

*Passage Excerpt*

**[P2]** Although airplanes and automobiles have now assumed greater prominence, the time has arrived for the resurgence of railroads. A revitalized and advanced railway system **would confer numerous essential benefits on both the United States and the globe**.

**[P4]** To invest in a modern network of railroads would improve the United States in much the same way that the first railroads did in the nineteenth and early twentieth centuries. A high-speed passenger rail system would dramatically transform American life as travel between cities and states became quicker and more convenient, **encouraging commerce, business, and tourism**. Such a system would **also make important strides in environmental preservation**. According to a 2007 British study, "CO2 emissions from aircraft operations are...at least five times greater" than those from high-speed trains. For similar reasons, Osaka, Japan, was ranked as "the best...green transportation city in Asia" by the 2011 *Green City Index*. As Lee-in Chen Chiu notes in *The Kyoto Economic Review*, Osakans travel by railway more than twice as much as they travel by car.

### Step 1: Restate the Relevant Claim or Argument

Looking at the passage, it is clear that the author is arguing in favor of building a high-speed rail system in America. This argument begins in Paragraph 2, and can be summed up as follows:

1. A new high-speed rail system would **benefit both the United States and the globe.**

2. These benefits include **economic gains** and better **environmental preservation.**

**[P5]** It is true that developing a countrywide high-speed rail system would come with significant costs. However, that was also true of the original Transcontinental Railroad, as indeed it is with virtually any great project undertaken for the public good. We should thus move ahead with confidence that **the rewards will outweigh the expenditure as citizens increasingly choose to travel by train**. Both for society's gain and the crucial well-being of the planet, **our path forward should proceed upon rails**.

3. Although it would be expensive, **the benefits will outweigh the costs as people choose to travel by train**.

4. Therefore, **we should build a high-speed rail system in America**.

### Step 2: Identify the Key Factors

Now that we have refamiliarized ourselves with the author's argument, we can see the key factors it depends on. Statements 1 and 4 are just general claims, while statements 2 and 3 give more details about the benefits of high-speed rail. Taking those statements together, the author argues:

**The economic and environmental benefits of high-speed rail will outweigh the costs as people choose to travel by train**.

Accordingly, the author's argument depends on the belief that Americans will choose to start traveling by high-speed rail once they have that option. But what if they didn't? If this assumption were wrong, none of the benefits of high-speed rail would be achieved. Therefore, the key factor in the author's argument is **the assumption that Americans will choose to travel by high-speed rail**.

**Step 3: Compare Answer Choices with the Key Factors**

We have determined that the author's argument works only if we **assume that Americans will choose to travel by high-speed rail**. Turning to the answer choices, **Choice D** expresses the same assumption in different words: *The public would readily adopt high-speed rail travel if it were available*. Therefore, this passage assertion is the most essential for readers to accept in order for them to find the author's argument reasonable.

Choice B may also seem like a plausible answer: if readers did not believe that *a high-speed railroad network would encourage economic activity*, they would not believe that building high-speed rail would lead to economic benefits. However, they might still find it reasonable to think high-speed rail would lead to *environmental* benefits. By contrast, without the belief that people will actually use high-speed rail, there would be no reason to think *any* benefits would be achieved. Therefore, while Choice B may initially seem attractive, **Choice D** is the best answer.


For an alternative method of explanation based more specifically on passage evidence, you can view this question in the UWorld Qbank (sold separately).

# Passage K: Lengthening the School Day

There may be reasons to reject the idea of lengthening the school day. None of them, however, are *good* reasons. Rather, the supposed demerits of such a proposal fall easily in the face of its numerous financial and social benefits for families.

The greatest of these benefits lies in reducing the need for childcare. It is a curious fact of American life that the adult's work schedule and the child's school schedule are misaligned. Children rise with the sun to head to classes, only to be sent home again hours before parents return from their jobs. In a society where, more often than not, both parents work, this discordance creates the need for an expensive arrangement to fill the gap in families' routines. For instance, studies show that in 2016, childcare costs accounted for 9.5 to 17.5 percent of median family income, depending on the state. Today, 40 percent of families nationwide spend over 15 percent of their income on childcare. Transportation to and from care sites only adds to that expense.

An additional advantage of an extended school day would be to allow for greater diversity and depth in curricula. Schools across the country have increasingly cut instruction in arts, music, and physical education (as well as recess) in order to meet objectives in reading and math. While this unfortunate state of affairs can be partially blamed on overzealous attention to standardized tests, it points to the larger deleterious trend of narrowing students' instruction. With a longer school day, such eliminated subjects can be restored, enriching students with a more well-rounded education.

To this proposal, however, critics may object that the added time would impose strain on educators. Can we truly ask schoolteachers—already among the most overworked individuals in society—to endure even more hours in the classroom? The answer is that a lengthened school day need not distress teachers nor add to their already cumbersome workload. By providing for additional areas of study in the arts and humanities, the extension would give schools cause to hire new, perhaps specialized, faculty to offer these courses. Moreover, the time could also be allocated to sports, academic clubs, and other extracurricular activities.

However, this point speaks to another objection, namely, the cost of adjusting the school day. Whether through paying current teachers more or hiring new ones, implementing such a proposal would entail a significant financial expenditure. There are at least two responses to this line of thinking. First is that this increase in the cost of schooling would be offset and likely surpassed by the aforementioned savings in childcare. Thus, while it is true that schools would require greater funding (likely necessitating higher property taxes), parents would ultimately pay the same or less overall, with greater educational opportunities for their children and fewer transportational burdens. Second is that schools should be better funded regardless. Recently, some schools—especially those in rural areas—have even reduced school weeks to only four days as a cost-saving measure. It is beyond dispute that schools across the board both need and deserve a radically increased investment from citizens. Lengthening the school day is simply one manifestation of how such funding should be utilized.

With this one change, states can coordinate the lives of parents and children, reduce the need for costly childcare, and expand curricular offerings. These worthy and desirable aims provide a clear justification for extending the school day.

*Lengthening the School Day* ©UWorld

Annotations for this passage can be found in the CARS Passage Booklet.

### 2a. Logical Relationships Within Passage Question 3

Which of the following pieces of information would be most important to know in order to assess the strength of the author's argument?

A. Whether the statistics on four-day school weeks represent a growing trend
B. Whether the statistics on childcare refer specifically to school-age children
C. What percentage of class time is currently spent on non-core subjects
D. What percentage of students currently participate in after-school activities

See next page for the *strategy-based explanation* of this question.

## 2a. Logical Relationships Within Passage Question 3
## Strategy-based Explanation

Which of the following pieces of information would be most important to know in order to assess the strength of the author's argument?

    A. Whether the statistics on four-day school weeks represent a growing trend
    B. Whether the statistics on childcare refer specifically to school-age children
    C. What percentage of class time is currently spent on non-core subjects
    D. What percentage of students currently participate in after-school activities

The question refers to "the author's argument" without providing any specifics. However, the whole passage is clearly an argument in favor of lengthening the school day. Therefore, we can think of the question as asking: "What do I most need to know to evaluate whether lengthening the school day is justified?"

## Applying the Method

*Passage Excerpt*

**[P1]** There may be reasons to reject the idea of **lengthening the school day**. None of them, however, are *good* reasons. Rather, the supposed demerits of such a proposal fall easily in the face of **its numerous financial and social benefits for families**.

**[P2] The greatest of these benefits lies in reducing the need for childcare**. It is a curious fact of American life that the adult's work schedule and the child's school schedule are misaligned. Children rise with the sun to head to classes, only to be sent home again hours before parents return from their jobs. In a society where, more often than not, both parents work, this discordance creates the need for an expensive arrangement to fill the gap in families' routines. For instance, studies show that in 2016, childcare costs accounted for 9.5 to 17.5 percent of median family income, depending on the state. Today, 40 percent of families nationwide spend over 15 percent of their income on childcare. Transportation to and from care sites only adds to that expense.

**Step 1: Restate the Relevant Claim or Argument**

The author argues that lengthening the school day would provide multiple benefits. More specifically, the argument progresses as follows:

1. **Lengthening the school day** would have **numerous financial and social benefits for families**.

2. **Reducing the need for childcare** is the greatest benefit.

**[P3] An additional advantage of an extended school day would be to allow for greater diversity and depth in curricula.** Schools across the country have increasingly cut instruction in arts, music, and physical education (as well as recess) in order to meet objectives in reading and math.

**[P4] To this proposal, however, critics may object** that the added time would impose strain on educators. Can we truly ask schoolteachers—already among the most overworked individuals in society—to endure even more hours in the classroom? The answer is that….

**[P5] However, this point speaks to another objection,** namely, the cost of adjusting the school day. Whether through paying current teachers more or hiring new ones, implementing such a proposal would entail a significant financial expenditure. There are at least two responses to this line of thinking.

**[P6]** With this one change, states **can coordinate the lives of parents and children, reduce the need for costly childcare, and expand curricular offerings.** These worthy and desirable aims **provide a clear justification for extending the school day**.

---

3. **Allowing a more diverse curriculum** is another benefit.

4. The author responds to **an objection**.

5. The author responds to **another objection**.

The author ends by summing up their argument:

6. We should lengthen the school day because it will **coordinate the lives of parents and children, reduce the need for costly childcare, and expand curricular offerings**.

## Step 2: Identify the Key Factors

We have broken the author's argument down into its main components. Thus, our next step is to identify the component most important for evaluating its strength. However, the author actually tells us what *they* think is the most important part of the argument:

2. **The greatest of these benefits lies in reducing the need for childcare.**

Accordingly, the author claims that the school day should be lengthened because of the benefits that change would provide, and they also claim that reducing the need for childcare is the most important of those benefits. Therefore, the strength of the author's argument would seem to depend most on **what they say about childcare**.

### Step 3: Compare Answer Choices with the Key Factors

Based on the author's own words, we concluded that **what they say about childcare** is the key to evaluating their argument. Reviewing the answer choices, the only one that concerns childcare is **Choice B**. Therefore, to assess the strength of the author's argument, it would be most important to know *whether the statistics on childcare refer specifically to school-age children*.

If these statistics do *not* refer to school-age children, there would be no reason to think childcare would be affected by a longer school day. In that case, the author's primary motivation for supporting a lengthened school day would be disproven. Consequently, this is the only answer that could call into question what the author identifies as **the greatest benefit** of lengthening the school day, making **Choice B** correct.


For an alternative method of explanation based more specifically on passage evidence, you can view this question in the UWorld Qbank (sold separately).

# Passage K: Lengthening the School Day

There may be reasons to reject the idea of lengthening the school day. None of them, however, are *good* reasons. Rather, the supposed demerits of such a proposal fall easily in the face of its numerous financial and social benefits for families.

The greatest of these benefits lies in reducing the need for childcare. It is a curious fact of American life that the adult's work schedule and the child's school schedule are misaligned. Children rise with the sun to head to classes, only to be sent home again hours before parents return from their jobs. In a society where, more often than not, both parents work, this discordance creates the need for an expensive arrangement to fill the gap in families' routines. For instance, studies show that in 2016, childcare costs accounted for 9.5 to 17.5 percent of median family income, depending on the state. Today, 40 percent of families nationwide spend over 15 percent of their income on childcare. Transportation to and from care sites only adds to that expense.

An additional advantage of an extended school day would be to allow for greater diversity and depth in curricula. Schools across the country have increasingly cut instruction in arts, music, and physical education (as well as recess) in order to meet objectives in reading and math. While this unfortunate state of affairs can be partially blamed on overzealous attention to standardized tests, it points to the larger deleterious trend of narrowing students' instruction. With a longer school day, such eliminated subjects can be restored, enriching students with a more well-rounded education.

To this proposal, however, critics may object that the added time would impose strain on educators. Can we truly ask schoolteachers—already among the most overworked individuals in society—to endure even more hours in the classroom? The answer is that a lengthened school day need not distress teachers nor add to their already cumbersome workload. By providing for additional areas of study in the arts and humanities, the extension would give schools cause to hire new, perhaps specialized, faculty to offer these courses. Moreover, the time could also be allocated to sports, academic clubs, and other extracurricular activities.

However, this point speaks to another objection, namely, the cost of adjusting the school day. Whether through paying current teachers more or hiring new ones, implementing such a proposal would entail a significant financial expenditure. There are at least two responses to this line of thinking. First is that this increase in the cost of schooling would be offset and likely surpassed by the aforementioned savings in childcare. Thus, while it is true that schools would require greater funding (likely necessitating higher property taxes), parents would ultimately pay the same or less overall, with greater educational opportunities for their children and fewer transportational burdens. Second is that schools should be better funded regardless. Recently, some schools—especially those in rural areas—have even reduced school weeks to only four days as a cost-saving measure. It is beyond dispute that schools across the board both need and deserve a radically increased investment from citizens. Lengthening the school day is simply one manifestation of how such funding should be utilized.

With this one change, states can coordinate the lives of parents and children, reduce the need for costly childcare, and expand curricular offerings. These worthy and desirable aims provide a clear justification for extending the school day.

*Lengthening the School Day* ©UWorld

Annotations for this passage can be found in the CARS Passage Booklet.

## 2a. Logical Relationships Within Passage Question 4

According to the author, the increase in the cost of schooling and associated higher property taxes that would be necessitated by a lengthened school day would be offset by an equal or greater savings in childcare costs. Which of the following would be the most reasonable objection to this claim?

A. The potential drawbacks to lengthening the school day are not all financial in nature.
B. The cost of childcare is too insignificant to offset the added expense.
C. The school day should ideally be lengthened without raising schools' financial expenditure.
D. The increased taxes would unfairly impact non-parents and businesses.

See next page for the *strategy-based explanation* of this question.

> ## 2a. Logical Relationships Within Passage Question 4
> ### Strategy-based Explanation
>
> According to the author, the increase in the cost of schooling and associated higher property taxes that would be necessitated by a lengthened school day would be offset by an equal or greater savings in childcare costs. Which of the following would be the most reasonable objection to this claim?
>
> A. The potential drawbacks to lengthening the school day are not all financial in nature.
> B. The cost of childcare is too insignificant to offset the added expense.
> C. The school day should ideally be lengthened without raising schools' financial expenditure.
> D. The increased taxes would unfairly impact non-parents and businesses.

Although this question is long, it is not as complex as it might appear. Essentially, it identifies a claim the author makes and asks what the best objection to that claim would be. Accordingly, we simply need to review that claim in the passage, and then think about how it might be challenged or questioned.

## Applying the Method

*Passage Excerpt*

**[P5]** However, this point speaks to another objection, namely, the cost of adjusting the school day. Whether through paying current teachers more or hiring new ones, implementing such a proposal would entail a significant financial expenditure. There are at least two responses to this line of thinking. First is that **this increase in the cost of schooling would be offset and likely surpassed by the aforementioned savings in childcare**. Thus, while it is true that schools would require greater funding (**likely necessitating higher property taxes**), **parents would ultimately pay the same or less overall**, with greater educational opportunities for their children and fewer transportational burdens.

**Step 1: Restate the Relevant Claim or Argument**

We can find the claim that the question refers to in Paragraph 5, as the author responds to concerns about the cost of lengthening the school day. Here the author makes the following argument:

1. Lengthening the school day will **probably require increased property taxes**.

2. However, **these increased taxes will be offset by reduced childcare expenses**.

3. Therefore, **parents will end up paying the same or less overall**.

The question focuses on Statement 2, but we have now placed it in relation to the rest of the author's argument.

## Step 2: Identify the Key Factors

We have restated the author's argument and noted its most relevant claim:

2. However, **these increased taxes will be offset by reduced childcare expenses**.

Since we know we're looking for an objection to this claim, let's take a moment to analyze it. If reduced childcare expenses would offset an increase in taxes, that would mean people who previously paid more for childcare would now pay less.

However, this change would have no effect on anyone who *doesn't* pay for childcare. For instance, non-parents would pay more in taxes but receive no offsetting reduction in expenses. Since these individuals would have no compensating benefit, they would only be harmed by the increased taxes. Therefore, one might reasonably object that lengthening the school day would help some people by disadvantaging others. Accordingly, a key factor in the author's claim that could warrant an objection is that **the increased taxes would be offset only for parents**.

## Step 3: Compare Answer Choices with the Key Factors

The author's claim was meant to show that the increased property taxes required for lengthening the school day would not ultimately be a drawback. As we have seen, however, a key factor in this claim is that the increased property taxes **would be offset specifically for parents** (but not for anyone else who must pay those taxes). Consequently, this increase would be a drawback for non-parents even if the author is correct about an offsetting reduction in childcare costs.

Looking at the answer choices, we see that **Choice D** reflects this idea: *The increased taxes would unfairly impact non-parents and businesses*. Therefore, this answer constitutes a reasonable objection to the author's claim and should be the correct answer.

For an alternative method of explanation based more specifically on passage evidence, you can view this question in the UWorld Qbank (sold separately).

Lesson 6.2

# RWT Subskill 2b. Function of Passage Claim

| Skill 2: Reasoning Within the Text ||
|---|---|
| **Subskill** | **Student Objective** |
| 2a. Logical Relationships Within Passage | Identify logical relationships between passage claims |
| **2b. Function of Passage Claim** | **Determine what function a particular claim serves in the passage** |
| 2c. Extent of Passage Evidence | Determine the extent to which the passage provides evidence for its claims |
| 2d. Connecting Claims With Evidence | Connect passage claims with supporting passage evidence |
| 2e. Determining Passage Perspectives | Determine the perspective of the author or another source in the passage |
| 2f. Drawing Additional Inferences | Draw additional inferences based on passage information |

To answer **2b. Function of Passage Claim** questions, you must determine a particular claim's purpose in the passage. For example, these questions may concern why the author includes a certain idea, what a statement is meant to accomplish, or how different sections of the passage relate to each other. A good way to think about these sorts of questions is that they all essentially ask: What was the author's reason for making *this* specific claim?

## Method for Answering 2b. Function of Passage Claim Questions

### Step 1: Review the Context in Which the Claim Occurs

The first step in determining a claim's function is to look at the context in which it appears. The connection between the claim and its surrounding sentences will show how it relates to the points being made in that section of the passage.

### Step 2: Consider the Progression of Ideas

The second step is to consider the progression of ideas in which the claim is included. What does the author say in leading up to the claim the question asks about? How does the claim help transition to the information that follows it? By noting this sequence of ideas, you will get a sense for how the claim contributes to a larger theme.

### Step 3: Ask What the Claim Adds

As with **1b. Meaning of Term** questions, a helpful last step is to ask what the claim adds. If this claim had *not* been made, what would be missing from the passage? By mentally removing the claim in question, you will likely notice a gap between passage ideas that were previously connected. Thus, the function of the claim is to bridge that gap in the progression of ideas, and the correct answer choice will describe this bridge.

> **Subskill Connection:**
> **1b. Meaning of Term and 2b. Function of Passage Claim**
>
> As Step 3 suggests, **2b. Function of Passage Claim** questions are in many ways like **1b. Meaning of Term** questions (see Lesson 5.2). To answer 1b questions you must recognize the role a term plays in the sentence or paragraph in which it appears. Similarly, to answer 2b questions you must recognize the role a claim plays in the section of the passage in which it appears. Thus, both types of questions ask how a component of the passage functions to help make a particular point.

## Subskill 2b. Function of Passage Claim
## (3 Practice Questions)

We now look at three examples from the UWorld Qbank. In each case, the passage and question are first given without commentary, allowing you to practice applying the method yourself. Then, the passage and question are presented again, this time with annotations of the passage (see the CARS Passage Booklet) and a step-by-step explanation of the question using the described method.

## Passage F: The Victorian Internet

Lasting roughly from 1820 to 1914, the Victorian Era is often defined by its many distinctive sociological conditions, including industrialization, urbanization, railroad travel, imperialism, territorial expansionism, and the frictions sparked by Darwinism and democratic reform. However, descriptors that may not come as readily to mind are "the Information Age" and "the Age of Communication"; nor would we likely associate this era with something called the "highway of thought," which emanated from the electric telegraph—an invention that, in retrospect, could be renamed "the Victorian Internet." Yet the rise of the electric telegraph in the mid-1800s constituted the greatest revolution in communication since the invention of the printing press in the fifteenth century and until the launch of the World Wide Web at the end of the twentieth century.

Historically, messages could be conveyed only as fast as a person could travel from one location to another. However, by the end of the eighteenth century, the Chappe brothers had constructed a rudimentary optical telegraph on a hilltop tower in France. Using the large, jointed arms attached to the roof, operators could form various configurations to communicate a message to a similar tower farther away, which would then relay the message to a third tower, and so on, in a long-distance chain. Nevertheless, even with telescopes, visibility severely hampered the efficacy of this semaphore system. After countless attempts by scientists to use electricity to transmit messages, innovators in both Britain and America, including the painter and polymath Samuel F.B. Morse, worked at harnessing electromagnetic forces to send communications via cable. But while in Britain the technology was initially reserved for railway signaling, in the U.S. it culminated in the transmission of a message along a 40-mile telegraphic wire from Washington, D.C., to Baltimore in 1844. Using a binary code of short and long electrical impulses, or "dots and dashes," Morse dispatched the words: "What hath God wrought?"

Within 20 years, telegraph cables crisscrossed the continental U.S. and much of Europe. The telegraph office became ubiquitous, and telegrams—often bouncing from one office to another like emails tossed from server to server—reached ever more remote destinations. Following some spectacular failures, a durable cable was stretched across the floor of the Atlantic Ocean, enabling Europe and the U.S. to exchange messages within minutes. In the 1870s, the British Empire connected London with outposts in India and Australia. While individual telegram speed remained relatively constant, an endless deluge of information began to pour through the wires. There was, as Tom Standage observes, "an irreversible acceleration in the pace of business life," reflecting how "telegraphy and commerce thrived in a virtuous circle." It was as though "rapid long-distance communication had effectively obliterated time and space," begetting the phenomenon known as "globalization."

A new type of skilled worker, the telegrapher, was born. He or she belonged to a vast, online community, whose semi-anonymous members shared a unique intermediary role as well as a language of dots and dashes—vaguely prefiguring the exchange of bits and bytes along modern computer networks. Like today's online communities, these telegraphers fostered their own subculture: jokes and anecdotes flew over the wires; some operators invented systems to play games; and occasional romances blossomed. Regrettably, hackers, scammers, and shady entrepreneurs also frequented the byways of this early internet.

Predictably, newer forms of the original telegraphic technology—first the telephone and, much later, the fax machine—eventually encroached. The last telegram departed from India in 2013, but the twenty-first century has arguably seen a revival of this communication form in the text message. As Standage claims, texting has not only "reincarnate[d] a defunct nineteenth-century technology" but reinforced "the democratization of telecommunications" inaugurated by the miraculous Victorian Internet.

*The Victorian Internet* ©UWorld

Annotations for this passage can be found in the CARS Passage Booklet.

### 2b. Function of Passage Claim Question 1

Why does the author most likely introduce the topic of the Victorian Internet by discussing the sociological conditions prevalent between 1820 and 1914?

- A. To explore the details of an era of change leading to the inevitable development of the telegraph
- B. To reveal the socioeconomic factors that made improvements in communication possible
- C. To suggest that the many dramatic changes of the period cause people to overlook the telegraph's impact
- D. To overcome the notion of the telegraph as a phenomenon that transcended its time

See next page for the *strategy-based explanation* of this question.

Chapter 6: Skill 2 – Reasoning Within the Text

> ## 2b. Function of Passage Claim Question 1
> ## Strategy-based Explanation
>
> Why does the author most likely introduce the topic of the Victorian Internet by discussing the sociological conditions prevalent between 1820 and 1914?
>
> A. To explore the details of an era of change leading to the inevitable development of the telegraph
> B. To reveal the socioeconomic factors that made improvements in communication possible
> C. To suggest that the many dramatic changes of the period cause people to overlook the telegraph's impact
> D. To overcome the notion of the telegraph as a phenomenon that transcended its time

This question asks you to consider the author's purpose in mentioning the sociological conditions of the times before going on to introduce its main topic, the telegraph or Victorian Internet. In other words, if the author's aim in the passage is to discuss the Victorian Internet, why do they bring up these other conditions first?

## Applying the Method

*Passage Excerpt*

**[P1]** Lasting roughly from 1820 to 1914, **the Victorian Era is often defined by its many distinctive sociological conditions,** including industrialization, urbanization, railroad travel, imperialism, territorial expansionism, and the frictions sparked by Darwinism and democratic reform. **However, descriptors that may not come as readily to mind** are "the Information Age" and "the Age of Communication"; nor would we likely associate this era with something called the "highway of thought," which emanated from **the electric telegraph—an invention that, in retrospect, could be renamed "the Victorian Internet."** Yet the rise of the electric telegraph in the mid-1800s constituted the greatest revolution in communication since the invention of the printing press in the fifteenth century and until the launch of the World Wide Web at the end of the twentieth century.

### Step 1: Review the Context in Which the Claim Occurs

In this case, the context for the claim is heavily suggested by the question itself. The passage begins by listing the **sociological conditions** typically thought to characterize the Victorian Era, such as **industrialization** and **urbanization**. The author then immediately follows this characterization by emphasizing a less recognized Victorian phenomenon, **the electric telegraph or "Victorian Internet."** Accordingly, the context for the claim is a discussion of the Victorian Era and its most significant features.

### Step 2: Consider the Progression of Ideas

The initial sentence of Paragraph 1 begins the following progression of ideas:

The author lists several sociological conditions **commonly associated with the Victorian Era.**

The author then highlights a different feature of the era **that may not come as readily to mind,** the effects of the **telegraph** on **information** and **communication.**

Finally, the author stresses the true significance of this phenomenon, calling **the rise of the**

> **telegraph the greatest revolution in communication** between the printing press and the World Wide Web.

### Step 3: Ask What the Claim Adds

As we have observed, the author contrasts many commonly recognized characteristics of the Victorian Era with the less recognized phenomenon of the telegraph. Without this introductory claim and the contrast it forms, the passage would still describe the telegraph's evolution, impact, and similarities with today's internet. However, the reader would have no indication that, despite its world-changing effects, the telegraph has often been **overshadowed** by the other features of the Victorian Era.

Accordingly, what the claim adds is the sense that the great significance of the telegraph, or "Victorian Internet," has been underacknowledged. Consequently, the reason the author introduces the Victorian Internet by discussing the other sociological conditions prevalent at the time is **Choice C**, *To suggest that the many dramatic changes of the period cause people to overlook the telegraph's impact*.

For an alternative method of explanation based more specifically on passage evidence, you can view this question in the UWorld Qbank (sold separately).

## Passage C: Probability and The Universe

The idea of probability is frequently misunderstood, in large part because of a conceptual confusion between objective probability and subjective probability. The failure to make this distinction leads to an erroneous conflation of genuine possibility with what is in fact merely personal ignorance of outcome. An example will clarify.

A standard die is rolled on a table, but the outcome of the roll is concealed. Should an observer be asked the chance that a particular number was rolled—five, say—the natural response is 1/6. However, this answer is incorrect. To say there is a one-in-six chance that a five was rolled implies there is an equal chance that any of the other numbers were rolled. But there is no equal chance, because the roll has already occurred. Hence, the probability that the result of the roll is a five is either 100% or 0%, and the same is true for each of the other numbers.

It might be objected that such an analysis is an issue of semantics rather than a substantive claim. For declaring the probability to be 1/6 is merely an expression that, for all we know, any number from 1 through 6 might have been rolled. But the difference between *for all we know* and what *is* remains crucially important, because it forestalls a tendency to make scientific assertions from a perspective biased by human perception….

[T]hus, it should be clear that claims about the purpose of the universe rest on shaky ground. In particular, we must be wary of inferences drawn from juxtaposing the existence of intelligent life with the genuinely improbable cosmological conditions that make such life possible. Joseph Zycinski provides an instructive account of those conditions: "If twenty billion years ago the rate of expansion were minimally smaller, the universe would have entered a stage of contraction at a time when temperatures of thousands of degrees were prevalent and life could not have appeared. If the distribution of matter were more uniform the galaxies could not have formed. If it were less uniform, instead of the presently observed stars, black holes would have formed from the collapsing matter." In short, the existence of life (let alone intelligent life) depends upon the initial conditions of the universe having conformed to an extremely narrow range of possible values.

It is tempting to go beyond Zycinski's factual point to draw deep cosmological, teleological, and even theological conclusions. But no such conclusions follow. The reason why a life-sustaining universe exists is that if it did not, there would be no one to wonder why a life-sustaining universe exists. This fact is a function not of purpose, but of pre-requisite. For instance, suppose again that a five was rolled on a die. We now observe a five on its face, not because the five was "meant to be" but because one side of the die had to land face up. Even if we imagine that the die has not six but six *billion* sides, that analysis is unaffected. Yes, it was unlikely that any given value would be rolled, but *some* value had to be rolled. That "initial condition" then set the parameters for what kind of events could possibly follow; in this case, the observation of the five.

Similarly, the existence of intelligent beings is evidence that certain physical laws obtained in the universe. However, it is not evidence that those beings or laws were necessary or intended rather than essentially random. The subjective probability of that outcome is irrelevant, because in this universe the objective probability of its occurrence is 100%. Thus, the seemingly low initial chance that such a universe would exist is not in itself indicative of a purpose to its existence.

*Probability and The Universe* ©UWorld

Annotations for this passage can be found in the CARS Passage Booklet.

## 2b. Function of Passage Claim Question 2

Why does the author ask readers to "imagine that the die has not six but six *billion* sides" (Paragraph 5)?

   A. To explain the tendency for people to confuse objective probability and subjective probability
   B. To illustrate that the improbability of an event does not imply that its occurrence has meaning
   C. To provide a counterpoint to the position advanced by Zycinski
   D. To show that there is no larger purpose to the existence of the universe

See next page for the *strategy-based explanation* of this question.

# Chapter 6: Skill 2 – Reasoning Within the Text

> ## 2b. Function of Passage Claim Question 2
> ## Strategy-based Explanation
>
> Why does the author ask readers to "imagine that the die has not six but six *billion* sides" (Paragraph 5)?
>
>    A. To explain the tendency for people to confuse objective probability and subjective probability
>    B. To illustrate that the improbability of an event does not imply that its occurrence has meaning
>    C. To provide a counterpoint to the position advanced by Zycinski
>    D. To show that there is no larger purpose to the existence of the universe

This question gives us a lot of information to work with. In addition to the paragraph reference, we can see that the quote referred to is based on a comparison between a normal die and one with six billion sides. Accordingly, we know exactly which part of the passage to review, and we know that we need to determine how this comparison impacts the claims being made there.

## Applying the Method

*Passage Excerpt*

**[P5]** It is tempting to go beyond Zycinski's factual point to draw deep cosmological, teleological, and even theological conclusions. But no such conclusions follow. The reason why a life-sustaining universe exists is that if it did not, there would be no one to wonder why a life-sustaining universe exists. This fact is a function not of purpose, but of pre-requisite. For instance, suppose again that a five was rolled on a die. We now observe a five on its face, not because the five was "meant to be" but because one side of the die had to land face up. **Even if we imagine that the die has not six but six *billion* sides, that analysis is unaffected.** Yes, it was unlikely that any given value would be rolled, but some value had to be rolled. That "initial condition" then set the parameters for what kind of events could possibly follow; in this case, the observation of the five.

### Step 1: Review the Context in Which the Claim Occurs

The quote in question comes as part of an example comparing the existence of a life-sustaining universe to the result of a die roll. According to the author, the die roll's outcome was not "meant to be"; it was just something that happened. The quote then expands on this example: **Even if we imagine that the die has not six but *six billion* sides, that analysis is unaffected.**

Therefore, the context in which this claim occurs is an example suggesting that the universe was not necessarily "meant to be."

### Step 2: Consider the Progression of Ideas

The quote comes near the end of the following sequence of ideas:

First, the author notes the temptation to go beyond the facts to draw conclusions about the universe, stating that no such conclusions follow.

The author then makes a distinction between the purpose of the universe's existence and the pre-requisite for the universe's existence.

Finally, the die-rolling example illustrates this distinction: **even if there were six billion**

> **possibilities**, some outcome would have to occur. Thus, there may be no *purpose* behind the "initial condition" of the universe.

### Step 3: Ask What the Claim Adds

As we have seen, the author uses a die-rolling example to illustrate why we should not draw conclusions about the universe that go beyond the facts. In this analogy, the fact that life-sustaining conditions obtained in the universe is like the fact that a certain number was rolled on a die: both may simply be random occurrences without any meaning or purpose behind them.

If the author did not ask readers to **imagine that the die has not six but six *billion* sides**, the overall point would still be the same: the existence of the universe was not "meant to be" simply because it was unlikely to have occurred. However, the inclusion of this statement amplifies that unlikelihood by several orders of magnitude. The author is essentially making the claim: "Even if there was only a one-in-six-billion chance that a life-sustaining universe would exist, that fact does not mean the universe has a purpose."

Accordingly, what the claim adds to the passage is the true scale of the author's position: that we cannot conclude the universe had a purpose *regardless* of how improbable its existence was. Therefore, the reason the author asks readers to **imagine that the die has not six but six *billion* sides** is Choice B, *To illustrate that the improbability of an event does not imply that its occurrence has meaning*.

For an alternative method of explanation based more specifically on passage evidence, you can view this question in the UWorld Qbank (sold separately).

## Passage G: Food Costs and Disease

Because frequent consumption of unhealthy foods is strongly linked with cardiometabolic diseases, one way for governments to combat those afflictions may be to modify the eating habits of the general public. Applying economic incentives or disincentives to various types of foods could potentially alter people's diets, leading to more positive health outcomes.

Utilizing national data from 2012 regarding food consumption, health, and economic status, Peñalvo et al. concluded that such price adjustments would help to prevent deaths related to cardiometabolic diseases. According to their analysis, increasing the prices of unhealthy foods such as processed meats and sugary sodas by 10%, while reducing the prices of healthy foods such as fruit and vegetables by 10%, would prevent an estimated 3.4% of yearly deaths in the U.S. Changing prices by 30% would have an even stronger effect, preventing an estimated 9.2% of yearly deaths. This data comports with that found in other countries, such as "previous modeling studies in South Africa and India, where a 20% SSB [sugar-sweetened beverage] tax was estimated to reduce diabetes prevalence by 4% over 20 years." The effects of price adjustments would be most pronounced on persons of lower socioeconomic status, as the researchers "found an overall 18.2% higher price-responsiveness for low versus high SES [socioeconomic status] groups."

This differential effect based on socioeconomic status contributes to concerns about such interventions, however. In *Harvard Public Health Review*, Kates and Hayward ask: "Well intentioned though they may be, at what point do these taxes overstep government influence on an individual's right to autonomy in decision-making? On whom does the increased financial burden of this taxation fall?" They note that taxes on sugar-sweetened beverages, for instance, "are likely to have a greater impact on low-income individuals…because individuals in those settings are more likely to be beholden to cost when making decisions about food."

However, "well intentioned though they may be," the worries that Kates and Hayward express are to some extent misguided. In particular, the idea that taxing unhealthy foods would burden those least able to afford it misses the point. Although the increased taxes would affect anyone who continued to purchase the items despite the higher prices, the goal of raising prices on unhealthy foods is precisely to dissuade people from buying them. As Kates and Hayward themselves remark, "Those in low-income environments may also be the largest consumers of obesogenic foods and therefore most likely to benefit from such a lifestyle change indirectly posed by SSB taxes." As the goal of the taxes is to promote those lifestyle changes, the financial burden objection is a non-starter.

Given this recognition, the question regarding autonomy constitutes a more substantial issue. Nevertheless, that concern also rests on a dubious assumption, as people's autonomy is not necessarily respected in the current situation either. The fact that those of lower socioeconomic status are more likely to have poorer diets suggests that such persons' food choices are the result of financial constraint, not fully autonomous, rational deliberation. Hence, by making healthy foods more affordable relative to unhealthy ones, government intervention might actually facilitate autonomous choices rather than hindering them.

On the other hand, suppose that the disproportionate consumption of unhealthy foods—and associated higher incidence of disease—among certain groups is not the result of financial hardship but rather the result of those persons' perceived self-interest. If so, that would suggest that members of these groups are being encouraged to persist in harmful dietary habits for the sake of corporate profits. In that case, violating autonomy for the sake of health may be permissible, as that practice would be morally preferable to the present system of corporate exploitation.

*Food Costs and Disease* ©UWorld

Annotations for this passage can be found in the CARS Passage Booklet.

## 2b. Function of Passage Claim Question 3

The passage author mentions some of Kates and Hayward's words again in Paragraph 4 most likely in order to:

A. illustrate an irony in these researchers' concerns about proposed price adjustments.
B. affirm these researchers' point that lower-income persons would benefit most from proposed price adjustments.
C. argue that social class is irrelevant to evaluating proposed price adjustments.
D. support these researchers' concerns about the effects of proposed price adjustments.

See next page for the *strategy-based explanation* of this question.

Chapter 6: Skill 2 – Reasoning Within the Text

> ## 2b. Function of Passage Claim Question 3
> ### Strategy-based Explanation
>
> The passage author mentions some of Kates and Hayward's words again in Paragraph 4 most likely in order to:
>
> A. illustrate an irony in these researchers' concerns about proposed price adjustments.
> B. affirm these researchers' point that lower-income persons would benefit most from proposed price adjustments.
> C. argue that social class is irrelevant to evaluating proposed price adjustments.
> D. support these researchers' concerns about the effects of proposed price adjustments.

The question asks why the author "mentions some of Kates and Hayward's words *again*." Looking at Paragraph 4, we can see that it contains two quotes from Kates and Hayward. However, the first quote repeats some of their words from Paragraph 3, so these must be the words the question refers to. Accordingly, our task is to determine the author's reason for repeating them.

## Applying the Method

*Passage Excerpt*

**[P3]** This differential effect based on socioeconomic status contributes to concerns about such interventions, however. In *Harvard Public Health Review*, Kates and Hayward ask: "**Well intentioned though they may be,** at what point do these taxes overstep government influence on an individual's right to autonomy in decision-making? On whom does the increased financial burden of this taxation fall?" They note that taxes on sugar-sweetened beverages, for instance, "are likely to have a greater impact on low-income individuals…because individuals in those settings are more likely to be beholden to cost when making decisions about food."

**[P4]** However, **"well intentioned though they may be,"** the worries that Kates and Hayward express are to some extent misguided. In particular, the idea that taxing unhealthy foods would burden those least able to afford it misses the point. Although the increased taxes would affect anyone who continued to purchase the items despite the higher prices, the goal of raising prices on unhealthy foods is precisely to dissuade people from buying them. As Kates and Hayward themselves

### Step 1: Review the Context in Which the Claim Occurs

As we have seen, the words that the author repeats in Paragraph 4 first appear in Paragraph 3.

Specifically, these words are part of Kates and Hayward's objection to food price interventions. **"Well intentioned though they may be,"** such interventions might create an unfair financial burden on persons of low socioeconomic status.

The author then repeats these words to introduce a response to Kates and Hayward's concerns in Paragraph 4. **"Well intentioned though they may be,"** these worries miss the point of the price interventions.

Thus, the context of the claim is the author's response or counterpoint to Kates and Hayward's concerns about food price interventions.

### Step 2: Consider the Progression of Ideas

The passage ideas relating to the quote and its repetition proceed as follows:

Kates and Hayward pose an objection to food price interventions. **(Paragraph 3)**

| | |
|---|---|
| remark, "Those in low-income environments may also be the largest consumers of obesogenic foods and therefore most likely to benefit from such a lifestyle change indirectly posed by SSB taxes." As the goal of the taxes is to promote those lifestyle changes, the financial burden objection is a non-starter. | The author responds to Kates and Hayward's objection, **repeating their own words** in the process. **(Paragraph 4)**<br><br>The author concludes that Kates and Hayward's objection fails. **(Paragraph 4)** |

### Step 3: Ask What the Claim Adds

If the author had *not* repeated Kates and Hayward's words, nothing would change about the progression of ideas in which those words currently appear. Paragraph 4 would simply begin: "However, the worries that Kates and Hayward express are to some extent misguided," without the interruption of the repeated quote.

Therefore, the only thing that repetition adds is the suggestion of a parallel between the author's response and Kates and Hayward's objection. Returning to Paragraph 3, Kates and Hayward claim there are problems with food price interventions, but they acknowledge that those interventions are based on worthy motivations. Thus, they declare that **"well intentioned though they may be,"** those price interventions are flawed.

Accordingly, by repeating these words, the author implies that this same point is true of Kates and Hayward's objection. Thus, the author declares that **"well intentioned though they may be,"** it is *Kates and Hayward's concerns* that are flawed. In this way, the author uses the researchers' own words against them.

Based on that usage, the answer that stands out is **Choice A**, *to illustrate an irony in these researchers' concerns about proposed price adjustments*. If Kates and Hayward's words apply to their own objection rather than to the price adjustments they criticized, that would constitute an ironic state of affairs. Consequently, the author most likely mentions these words again to highlight that irony.

For an alternative method of explanation based more specifically on passage evidence, you can view this question in the UWorld Qbank (sold separately).

Lesson 6.3

# RWT Subskill 2c. Extent of Passage Evidence

| Skill 2: Reasoning Within the Text ||
| --- | --- |
| **Subskill** | **Student Objective** |
| 2a. Logical Relationships Within Passage | Identify logical relationships between passage claims |
| 2b. Function of Passage Claim | Determine what function a particular claim serves in the passage |
| **2c. Extent of Passage Evidence** | **Determine the extent to which the passage provides evidence for its claims** |
| 2d. Connecting Claims With Evidence | Connect passage claims with supporting passage evidence |
| 2e. Determining Passage Perspectives | Determine the perspective of the author or another source in the passage |
| 2f. Drawing Additional Inferences | Draw additional inferences based on passage information |

There are two RWT subskills that focus on passage evidence. The first of these subskills, **2c. Extent of Passage Evidence**, looks at whether the passage provides supporting evidence for its claims. These kinds of questions can be asked in various ways; for example, you might encounter any of the following phrasings:

- Which of the following statements is most supported by evidence or examples in the passage?
- The author does NOT provide any information to help explain:
- For which of the following passage claims does the author provide the most supporting evidence?
- Which of the following passage claims is LEAST supported by examples in the passage?

Whatever the wording, these questions relate to the quantity of evidence, not its quality. In other words, they do not ask you to *evaluate* the support given for a claim, only to determine whether support was provided. Accordingly, this subskill involves distinguishing statements that are merely asserted from those that are backed up by evidence.

Chapter 6: Skill 2 – Reasoning Within the Text

### What Counts as Evidence?

As a scientist, you may think of evidence in the strict sense of observations and data. For CARS, however, evidence is a broader concept that can also include theoretical arguments and hypothetical considerations.

If these seem like odd things to call "evidence," keep in mind that many CARS passages are concerned with conceptual claims or how certain ideas should be understood. For instance, an author might pose a thought experiment, then argue that a real-life situation is analogous and should be judged the same way. From that standpoint, it makes sense that the thought experiment could serve as evidence for the author's point.

Ultimately, you can think of evidence in CARS as **any additional reason the passage provides to think that a statement is true.**

When approaching these questions, remember that in most cases a statement is not evidence *for itself*. Occasionally students are confused when they miss this type of question, asking: "This answer choice is directly stated in the passage, so how can you say there's no evidence for it?" But that is exactly the sort of distinction the question is asking you to see: making a statement is not the same as providing evidence for that statement. As mentioned above, you are looking for *additional* information that *supports* the original statement.

## Method for Answering 2c. Extent of Passage Evidence Questions

### Step 1: Review the Relevant Claims in the Passage

The first step is to review each answer choice and see what else is said where that claim appears in the passage. For instance, does the author mention additional points related to the claim, or do they just assert it and move on? Was the claim introduced by the preceding discussion, or does it represent a new point? In many (though not all) cases, the text immediately before or after a claim is where its evidence will appear.

### Step 2: Ask "What Reason Is Given to Believe This Claim?"

Next, for each claim you should ask some version of the question "What reason is given to believe this?" If a claim is supported by evidence, there will be specific information you can point to in the passage that tells you *why* you should accept the claim as true. If there is no such information, then the claim is not supported by evidence.

**Step 3: Match Claims to Evidence**

Finally, you will have identified the specific information that supports some claims, and found that other claims have been asserted without evidence. Depending on what the question asks for, the correct answer will be either the only statement that is supported or the only one that is not.

Note that it is also a good habit to follow Step 2 as a general practice. If you approach a passage with the mindset of asking why you should believe its claims, you will understand the passage better and prepare yourself to more easily answer its questions.

## Subskill 2c. Extent of Passage Evidence
## (3 Practice Questions)

We now look at three examples from the UWorld Qbank. In each case, the passage and question are first given without commentary, allowing you to practice applying the method yourself. Then, the passage and question are presented again, this time with annotations of the passage (see the CARS Passage Booklet) and a step-by-step explanation of the question using the described method.

# Passage D: American Local Motives

Locomotives were invented in England, with the first major railroad connecting Liverpool and Manchester in 1830. However, it was in America that railroads would be put to the greatest use in the nineteenth century. On May 10, 1869, the Union Pacific and Central Pacific lines met at Promontory Point, Utah, joining from opposite directions to complete a years-long project—the Transcontinental Railroad. This momentous event connected the eastern half of the United States with its western frontier and facilitated the construction of additional lines in between. As a result, journeys that had previously taken several months by horse and carriage now required less than a week's travel. By 1887 there were nearly 164,000 miles of railroad tracks in America, and by 1916 that number had swelled to over 254,000.

While the United States still has the largest railroad network in the world, it operates largely in the background of American life, and citizens no longer view trains with the sense of importance those machines once commanded. Nevertheless, the economic and industrial advantages those citizens enjoy today would not have been possible without America's history of trains; as Tom Zoellner reminds us, "Under the skin of modernity lies a skeleton of railroad tracks." Although airplanes and automobiles have now assumed greater prominence, the time has arrived for the resurgence of railroads. A revitalized and advanced railway system would confer numerous essential benefits on both the United States and the globe.

The chief obstacles to garnering support for such a project are the current dominance of the automobile and the languishing technology of existing railroads. In a sense these two obstacles are one, as American dependence on personal automobiles is partially due to the paucity of rapid public transportation. The railroads of Europe and Japan, by comparison, have vastly outpaced their American counterparts. Japan has operated high-speed rail lines continuously since 1964, and in 2007, a French train set a record of 357 miles per hour. While that speed was achieved under tightly controlled conditions, it still speaks to the great disparity in railroad development between the United States and other countries since the mid-twentieth century. British trains travel at speeds much higher than those in America, where both the trains themselves and the infrastructure to support them have simply been allowed to fall behind. In much of Europe it is common for trains to travel at close to 200 miles per hour.

To invest in a modern network of railroads would improve the United States in much the same way that the first railroads did in the nineteenth and early twentieth centuries. A high-speed passenger rail system would dramatically transform American life as travel between cities and states became quicker and more convenient, encouraging commerce, business, and tourism. Such a system would also make important strides in environmental preservation. According to a 2007 British study, "CO2 emissions from aircraft operations are...at least five times greater" than those from high-speed trains. For similar reasons, Osaka, Japan, was ranked as "the best...green transportation city in Asia" by the 2011 *Green City Index*. As Lee-in Chen Chiu notes in *The Kyoto Economic Review*, Osakans travel by railway more than twice as much as they travel by car.

It is true that developing a countrywide high-speed rail system would come with significant costs. However, that was also true of the original Transcontinental Railroad, as indeed it is with virtually any great project undertaken for the public good. We should thus move ahead with confidence that the rewards will outweigh the expenditure as citizens increasingly choose to travel by train. Both for society's gain and the crucial well-being of the planet, our path forward should proceed upon rails.

*American Local Motives* ©UWorld

Annotations for this passage can be found in the CARS Passage Booklet.

## 2c. Extent of Passage Evidence Question 1

For which of the following passage claims does the author provide supporting evidence?

   A. Developing a countrywide high-speed rail system would come with significant costs.
   B. A high-speed passenger rail system would dramatically transform American life.
   C. The United States still has the largest railroad network in the world.
   D. Increasing the use of high-speed railroads would be beneficial for the environment.

See next page for the *strategy-based explanation* of this question.

## 2c. Extent of Passage Evidence Question 1
## Strategy-based Explanation

For which of the following passage claims does the author provide supporting evidence?

A. Developing a countrywide high-speed rail system would come with significant costs.
B. A high-speed passenger rail system would dramatically transform American life.
C. The United States still has the largest railroad network in the world.
D. Increasing the use of high-speed railroads would be beneficial for the environment.

Given how the question is phrased, we know that only one answer choice is supported in the passage. In other words, while three of the choices are claims asserted without evidence, the passage gives a specific reason to support the correct answer. Therefore, we are looking for that one supported answer choice.

## Applying the Method

*Passage Excerpt*

[P2] While the United States still has the largest railroad network in the world, it operates largely in the background of American life, and citizens no longer view trains with the sense of importance those machines once commanded.

[P4] To invest in a modern network of railroads would improve the United States in much the same way that the first railroads did in the nineteenth and early twentieth centuries. A high-speed passenger rail system would dramatically transform American life as travel between cities and states became quicker and more convenient, encouraging commerce, business, and tourism. **Such a system would also make important strides in environmental preservation. According to a 2007 British study,** "CO2 emissions from aircraft operations are...at least five times greater" than those from high-speed trains. For similar reasons, Osaka, Japan, was ranked as "the best...green transportation city in Asia" by **the 2011 *Green City Index*.** As Lee-in Chen Chiu notes in *The Kyoto Economic Review*, Osakans travel by railway more than twice as much as they travel by car.

**Step 1: Review the Relevant Claims in the Passage**

**Step 2: Ask "What Reason Is Given to Believe This Claim?"**

Choice C

"The United States still has the largest railroad network in the world"

- Paragraph 2 includes the Choice C claim that the United States still has the largest railroad network in the world.

Reasons Given to Believe This Claim

- However, the passage says nothing more about the size of the U.S. railroad network.

Choice B

"A high-speed passenger rail system would dramatically transform American life."

- Paragraph 4 states that high-speed rail would dramatically transform American life in terms of travel, commerce, business, and tourism.

Reasons Given to Believe This Claim

- However, no reason is given to believe that such a dramatic transformation will occur.

[P5] It is true that developing a countrywide high-speed rail system would come with significant costs. However, that was also true of the original Transcontinental Railroad, as indeed it is with virtually any great project undertaken for the public good. We should thus move ahead with confidence that the rewards will outweigh the expenditure as citizens increasingly choose to travel by train. Both for society's gain and the crucial well-being of the planet, our path forward should proceed upon rails.

**Choice D**

"Increasing the use of high-speed railroads would be beneficial for the environment."

- Paragraph 4 says high-speed rail would make **important strides in environmental preservation**.

**Reasons Given to Believe This Claim**

- Immediately after claiming that high-speed rail **helps preserve the environment**, the author provides **sets of data** to explain why this claim should be believed: studies showing that high-speed rail is greener than both airplanes and cars.

**Choice A**

"Developing a countrywide high-speed rail system would come with significant costs."

- Paragraph 5 talks about the **significant costs** that developing high-speed rail would involve.

**Reasons Given to Believe This Claim**

- However, the author provides no information about such costs or why we should believe they would be significant.

### Step 3: Match Claims to Evidence

By reviewing each answer choice in the context of the passage, we saw that only one of them includes a reason to think it is true. The author's claims about the environmental benefit of high-speed rail are supported by **a 2007 British study** comparing CO2 emissions of aircraft and high-speed trains and the **2011 *Green City Index*** comparing the environmental impact of railways and cars. Accordingly, the author provides this data as evidence for **Choice D**, *Increasing the use of high-speed railroads would be beneficial for the environment*.

We might have been tempted to think Choice B could be correct, since the author does elaborate on how *a high-speed passenger rail system would dramatically transform American life.* However, describing the impact on American life is part of the author's claim about a dramatic transformation; it is not itself a reason to believe such a transformation would take place. By contrast, **Choice D** is clearly supported by evidence.

For an alternative method of explanation based more specifically on passage evidence, you can view this question in the UWorld Qbank (sold separately).

# Passage J: The Inkblots

For almost 100 years now, the psychological evaluation known as the Rorschach Inkblot Test has engendered much controversy, including skepticism about its value, questions about its scoring, and, especially, criticism of its interpretive methods as too subjective. Thus, the Rorschach test, which emerged from the same early twentieth-century zeitgeist that produced Einstein's physics, Freudian psychoanalysis, and abstract art, seems one of modernity's most misbegotten children. Destined never to be completely accepted or discredited, the test remains a perennial outlier in its field. Nevertheless, the inkblots' mystery and aesthetic appeal have caused them to be indelibly printed on our cultural fabric.

The now iconic inkblots were introduced to the world by Swiss psychiatrist Hermann Rorschach in his 1921 book *Psychodiagnostics*. As both director of the Herisau Asylum in Switzerland and an amateur artist, Rorschach was uniquely positioned to wed the new practice of psychoanalysis to the budding phenomenon of abstract art. For instance, reading Freud's work on dream symbolism prompted him to recall his childhood passion for a game based on inkblot art called *Klecksographie*. He was also cognizant that in a recently published dissertation, his colleague Szymon Hens had used inkblots to try to probe the imagination of research subjects; moreover, a few years earlier, the French psychologist and father of intelligence testing Alfred Binet had used them to measure creativity.

Motivated by these developments, the Herisau director decided to revisit that childhood pastime that had awakened his curiosity about how visual information is processed. In particular, he wondered why different people saw different things in the same image. Traditionally, psychoanalysts had relied on language for insights; however, as biographer Damion Searls reports, Rorschach's theories would exemplify the principle that "who we are is a matter less of what we say than of what we see." Indeed, through a process of perception termed pareidolia, the mind projects meaning onto images, detecting in them familiar objects or shapes. Consequently, what a person sees in an image reveals more about that person than about the image itself.

Rorschach experimented with countless inkblots, eventually selecting ten—five black on white, two also featuring some red, and three pastel-colored—to use with research subjects. For these perfectly symmetrical images—each of which he was said to have "meticulously designed to be as ambiguous and 'conflicted' as possible"—the primary question was always "What do *you* see?" Rorschach was especially careful to note how much attention individuals paid to various components of each inkblot (such as form, color, and a sense of movement) and whether they concentrated on details or the whole image. Having observed that his patients with schizophrenia gave distinctly different responses from the control group, Rorschach envisioned his experiment as a diagnostic tool for the disease. Nevertheless, he resisted the notion that its results could be used to assess personality. In fact, until his untimely death from a ruptured appendix in 1922, Rorschach referred to his project as an "interpretive form experiment" rather than a test. Ironically, however, by the 1960s, the Rorschach Inkblot Test was known chiefly as a personality assessment and had become the most frequently administered projective personality test in the US.

Rorschach's test has survived nearly incessant scrutiny, including a 2013 comprehensive study of all Rorschach test data and repeated revisions to its scoring, yet doubts about its validity and reliability persist. Much like the inkblots themselves—which tantalize us with the possibility of divulging the secrets of who we are and how we see the world—the test has (for better or worse) defied attempts to fix its meaning. Thus, what has been called "the twentieth century's most visionary synthesis of art and science" stands tempered by harsh criticism.

*The Inkblots* ©UWorld

Annotations for this passage can be found in the CARS Passage Booklet.

## 2c. Extent of Passage Evidence Question 2

Which of the following passage claims is the LEAST supported by passage evidence?

- A. Rorschach's work remained a perennial outlier in its field.
- B. Rorschach's inkblots were conspicuous for their ambiguity.
- C. Rorschach's test emerged from the same zeitgeist that produced Einstein's physics.
- D. Rorschach's research involved the process of perception called pareidolia.

See next page for the *strategy-based explanation* of this question.

## 2c. Extent of Passage Evidence Question 2
### Strategy-Based Explanation

Which of the following passage claims is the LEAST supported by passage evidence?

A. Rorschach's work remained a perennial outlier in its field.
B. Rorschach's inkblots were conspicuous for their ambiguity.
C. Rorschach's test emerged from the same zeitgeist that produced Einstein's physics.
D. Rorschach's research involved the process of perception called pareidolia.

Based on the phrasing of the question, we can infer that three of the answer choices are supported by evidence while one answer choice is not. Accordingly, we are looking for the answer that is merely asserted but not further discussed in the passage.

## Applying the Method

*Passage Excerpt*

**[P1]** For almost 100 years now, the psychological evaluation known as the Rorschach Inkblot Test has engendered much controversy, including skepticism about its value, questions about its scoring, and, especially, criticism of its interpretive methods as too subjective. Thus, **the Rorschach test**, which **emerged from the same early twentieth-century zeitgeist that produced Einstein's physics**, Freudian psychoanalysis, and abstract art, seems one of modernity's most misbegotten children. Destined never to be completely accepted or discredited, **the test remains a perennial outlier in its field**. Nevertheless, the inkblots' mystery and aesthetic appeal have caused them to be indelibly printed on our cultural fabric.

**[P3]** Motivated by these developments, the Herisau director decided to revisit that childhood pastime that had awakened his curiosity about how visual information is processed. In particular, he wondered why different people saw different things in the same image. Traditionally, psychoanalysts had relied on language for insights; however, as biographer Damion Searls reports, Rorschach's theories would exemplify the principle that "who we are is a matter less of what we say than of what we see." **Indeed, through a process of perception termed pareidolia**, the mind projects meaning onto images,

### Step 1: Review the Relevant Claims in the Passage

### Step 2: Ask "What Reason Is Given to Believe This Claim?"

**Choice C**

"Rorschach's test emerged from the same zeitgeist that produced Einstein's physics."

- Paragraph 1 includes the Choice C statement that **Rorschach's test emerged from the same zeitgeist that produced Einstein's physics**, along with Freudian psychoanalysis and abstract art.

**Reasons Given to Believe This Claim**

- The author never elaborates on how or why the Rorschach test and Einstein's physics developed out of the same cultural context.

**Choice A**

"Rorschach's work remained a perennial outlier in its field."

- Paragraph 1 conveys the Choice A statement that **Rorschach's work remained a perennial outlier in its field**.

**Reasons Given to Believe This Claim**

- The passage suggests a few reasons why the Rorschach test was considered an outlier. For instance, it produced considerable controversy, particularly because its interpretive methods seemed too subjective. The author emphasizes that the test was "destined never to be completely accepted or discredited."

detecting in them familiar objects or shapes. Consequently, what a person sees in an image reveals more about that person than about the image itself.

[P4] Rorschach experimented with countless **inkblots**, eventually selecting ten—five black on white, two also featuring some red, and three pastel-colored—to use with research subjects. For these perfectly symmetrical images—each of which he was said to have "**meticulously designed to be as ambiguous** and 'conflicted' **as possible**"—the primary question was always "What do *you* see?" Rorschach was especially careful to note how much attention individuals paid to various components of each inkblot (such as form, color, and a sense of movement) and whether they concentrated on details or the whole image. Having observed that his patients with schizophrenia gave distinctly different responses from the control group, Rorschach envisioned his experiment as a diagnostic tool for the disease. Nevertheless, he resisted the notion that its results could be used to assess personality. In fact, until his untimely death from a ruptured appendix in 1922, Rorschach referred to his project as an "interpretive form experiment" rather than a test. Ironically, however, by the 1960s, the Rorschach Inkblot Test was known chiefly as a personality assessment and had become the most frequently administered projective personality test in the US.

[P5] Rorschach's test has survived nearly incessant scrutiny, including a 2013 comprehensive study of all Rorschach test data and repeated revisions to its scoring, yet doubts about its validity and reliability persist. Much like the inkblots themselves—which tantalize us with the possibility of divulging the secrets of who we are and how we see the world—the test has (for better or worse) defied attempts to fix its meaning. Thus, what has been called "the twentieth century's most visionary synthesis of art and science" stands tempered by harsh criticism.

In addition, Paragraph 5 returns to this theme by noting that the test's validity and reliability are still doubted and that its meaning has never been fixed.

### Choice D

"Rorschach's research involved the process of perception called pareidolia."

- In describing how the Rorschach test works, Paragraph 3 implies the Choice D claim that **Rorschach's research involved the process of perception called pareidolia**.

### Reasons Given to Believe This Claim

- According to Paragraph 3, the Rorschach test is based on the fact that different people see different things in the same inkblots. The author then indicates that the psychological phenomenon of **pareidolia** describes this process more generally: "the mind projects meaning onto images, detecting in them familiar objects or shapes." Thus, the author conveys that pareidolia is the reason why, in the Rorschach test, "what a person sees in an image reveals more about that person than about the image itself."

### Choice B

"Rorschach's inkblots were conspicuous for their ambiguity."

- In describing the actual inkblots, which were **"meticulously designed to be as ambiguous** and 'conflicted' **as possible**," Paragraph 4 makes a claim similar to that of Choice B.

### Reasons Given to Believe This Claim

- In Paragraph 4, the author elaborates on the ambiguity of the inkblots.

    The author describes how "the primary question for the inkblots was always 'What do *you* see?'" The implication is that the meaning of the inkblots was not fixed but varied according to each individual tested.

    The passage then describes some ways that responses to the inkblots varied. For instance, Rorschach noted differences in "how much attention individuals paid to various components of each inkblot (such as form, color, and a sense of movement) and whether they concentrated on details or the whole image." That test-takers have different reactions to the same images reinforces the idea that the inkblots were inherently ambiguous.

### Step 3: Match Claims to Evidence

By looking at each answer choice in the context of the passage, we can determine that **Choice C**, *Rorschach's test emerged from the same zeitgeist that produced Einstein's physics*, is not backed up by evidence. Although this claim is asserted, the author provides no further explanation or reason to think that it is true. By contrast, the other answer choices are all supported by additional information in the passage. Accordingly, **Choice C** is the correct answer.

For an alternative method of explanation based more specifically on passage evidence, you can view this question in the UWorld Qbank (sold separately).

## Passage K: Lengthening the School Day

There may be reasons to reject the idea of lengthening the school day. None of them, however, are *good* reasons. Rather, the supposed demerits of such a proposal fall easily in the face of its numerous financial and social benefits for families.

The greatest of these benefits lies in reducing the need for childcare. It is a curious fact of American life that the adult's work schedule and the child's school schedule are misaligned. Children rise with the sun to head to classes, only to be sent home again hours before parents return from their jobs. In a society where, more often than not, both parents work, this discordance creates the need for an expensive arrangement to fill the gap in families' routines. For instance, studies show that in 2016, childcare costs accounted for 9.5 to 17.5 percent of median family income, depending on the state. Today, 40 percent of families nationwide spend over 15 percent of their income on childcare. Transportation to and from care sites only adds to that expense.

An additional advantage of an extended school day would be to allow for greater diversity and depth in curricula. Schools across the country have increasingly cut instruction in arts, music, and physical education (as well as recess) in order to meet objectives in reading and math. While this unfortunate state of affairs can be partially blamed on overzealous attention to standardized tests, it points to the larger deleterious trend of narrowing students' instruction. With a longer school day, such eliminated subjects can be restored, enriching students with a more well-rounded education.

To this proposal, however, critics may object that the added time would impose strain on educators. Can we truly ask schoolteachers—already among the most overworked individuals in society—to endure even more hours in the classroom? The answer is that a lengthened school day need not distress teachers nor add to their already cumbersome workload. By providing for additional areas of study in the arts and humanities, the extension would give schools cause to hire new, perhaps specialized, faculty to offer these courses. Moreover, the time could also be allocated to sports, academic clubs, and other extracurricular activities.

However, this point speaks to another objection, namely, the cost of adjusting the school day. Whether through paying current teachers more or hiring new ones, implementing such a proposal would entail a significant financial expenditure. There are at least two responses to this line of thinking. First is that this increase in the cost of schooling would be offset and likely surpassed by the aforementioned savings in childcare. Thus, while it is true that schools would require greater funding (likely necessitating higher property taxes), parents would ultimately pay the same or less overall, with greater educational opportunities for their children and fewer transportational burdens. Second is that schools should be better funded regardless. Recently, some schools—especially those in rural areas—have even reduced school weeks to only four days as a cost-saving measure. It is beyond dispute that schools across the board both need and deserve a radically increased investment from citizens. Lengthening the school day is simply one manifestation of how such funding should be utilized.

With this one change, states can coordinate the lives of parents and children, reduce the need for costly childcare, and expand curricular offerings. These worthy and desirable aims provide a clear justification for extending the school day.

*Lengthening the School Day* ©UWorld

Annotations for this passage can be found in the CARS Passage Booklet.

## 2c. Extent of Passage Evidence Question 3

For which of the following assertions does the author provide the most support?

- A. Schools pay too much attention to standardized tests.
- B. It is beyond dispute that schools should receive more funding.
- C. Schoolteachers are some of the most overworked individuals in society.
- D. Reasons to maintain the current school day's duration are easily refuted.

See next page for the *strategy-based explanation* of this question.

## 2c. Extent of Passage Evidence Question 3
### Strategy-Based Explanation

For which of the following assertions does the author provide the most support?

- A. Schools pay too much attention to standardized tests.
- B. It is beyond dispute that schools should receive more funding.
- C. Schoolteachers are some of the most overworked individuals in society.
- D. Reasons to maintain the current school day's duration are easily refuted.

The wording of the question sounds as if it's suggesting that some answer choices are supported more than others. Typically, however, these questions just distinguish whether claims are supported or unsupported. Accordingly, we are most likely looking for the one answer choice that is backed up by passage evidence.

## Applying the Method

*Passage Excerpt*

**[P1]** There may be reasons to reject the idea of lengthening the school day. None of them, however, are *good* reasons. Rather, **the supposed demerits of such a proposal fall easily** in the face of its numerous financial and social benefits for families.

**[P3]** Schools across the country have increasingly cut instruction in arts, music, and physical education (as well as recess) in order to meet objectives in reading and math. While this unfortunate state of affairs can be partially blamed on overzealous attention to standardized tests, it points to the larger deleterious trend of narrowing students' instruction. With a longer school day, such eliminated subjects can be restored, enriching students with a more well-rounded education.

**[P4]** To this proposal, however, **critics may object that** the added time would impose strain on educators. Can we truly ask schoolteachers—already among the most overworked individuals in society—to endure even more hours in the classroom? **The answer is that** a lengthened school day need not distress teachers nor add to their already cumbersome workload. By providing for additional areas of study in the arts and humanities, the extension would give schools cause to hire

### Step 1: Review the Relevant Claims in the Passage

### Step 2: Ask "What Reason Is Given to Believe This Claim?"

**Choice D**

"Reasons to maintain the current school day's duration are easily refuted."

- Paragraph 1 states that the supposed demerits of lengthening the school day fall easily.

**Reasons Given to Believe This Claim**

- The author later mentions some objections to lengthening the school day, and provides a counterpoint for each.

    Paragraph 4 describes what **critics may object** regarding the potential **strain on educators**.

    The author then explains what they see as **the answer** to that concern.

    Paragraph 5 raises **another objection**, namely, the cost of adjusting the school day.

    The author replies that **there are at least two responses** to this line of thinking, and then goes on to describe both responses.

**Choice A**

"Schools pay too much attention to standardized tests."

- Paragraph 3 refers to what the author calls overzealous attention to standardized tests, implying that schools pay too much attention to these tests.

new, perhaps specialized, faculty to offer these courses. Moreover, the time could also be allocated to sports, academic clubs, and other extracurricular activities.

[P5] However, this point speaks to **another objection**, namely, the cost of adjusting the school day. Whether through paying current teachers more or hiring new ones, implementing such a proposal would entail a significant financial expenditure. **There are at least two responses** to this line of thinking. **First is that** this increase in the cost of schooling would be offset and likely surpassed by the aforementioned savings in childcare. Thus, while it is true that schools would require greater funding (likely necessitating higher property taxes), parents would ultimately pay the same or less overall, with greater educational opportunities for their children and fewer transportational burdens. **Second is that** schools should be better funded regardless. Recently, some schools—especially those in rural areas—have even reduced school weeks to only four days as a cost-saving measure. It is beyond dispute that schools across the board both need and deserve a radically increased investment from citizens. Lengthening the school day is simply one manifestation of how such funding should be utilized.

Reasons Given to Believe This Claim

- However, the passage says nothing else about standardized testing.

Choice C

"Schoolteachers are some of the most overworked individuals in society."

- In Paragraph 4, the author describes teachers as among the most overworked individuals in society.

Reasons Given to Believe This Claim

- The author does not provide any statistics or elaborating information to show that teachers are overworked.

Choice B

"It is beyond dispute that schools should receive more funding."

- Paragraph 5 includes the author's assertion: It is beyond dispute that schools across the board both need and deserve a radically increased investment from citizens.

Reasons Given to Believe This Claim

Although the author states that this assertion "is beyond dispute," they offer no reasons why someone could not dispute it. Similarly, the author provides no reasons why schools specifically *deserve* an increased investment, but seems to presume that the reader will agree with this claim.

**Step 3: Match Claims to Evidence**

We have seen that three of the answer choices are stated in the passage but are not supported by evidence. However, **Choice D**, *Reasons to maintain the current school day's duration are easily refuted*, is supported by information in Paragraphs 4 and 5.

While **critics may object** that lengthening the school day would impose strain on educators, the author offers the **counter-objection** that new educators would be hired to cover the extra work. In reply to **another objection** about the cost of adjusting the school day, the author provides **two responses:** that there will be a compensating reduction in childcare costs, and that schools should receive more funding anyway.

Accordingly, the author provides refutations to the claim that the school day should stay the same length. Therefore, **Choice D** is supported by evidence and is the correct answer.

For an alternative method of explanation based more specifically on passage evidence, you can view this question in the UWorld Qbank (sold separately).

Lesson 6.4

# RWT Subskill 2d. Connecting Claims With Evidence

| Skill 2: Reasoning Within the Text ||
|---|---|
| **Subskill** | **Student Objective** |
| 2a. Logical Relationships Within Passage | Identify logical relationships between passage claims |
| 2b. Function of Passage Claim | Determine what function a particular claim serves in the passage |
| 2c. Extent of Passage Evidence | Determine the extent to which the passage provides evidence for its claims |
| **2d. Connecting Claims With Evidence** | **Connect passage claims with supporting passage evidence** |
| 2e. Determining Passage Perspectives | Determine the perspective of the author or another source in the passage |
| 2f. Drawing Additional Inferences | Draw additional inferences based on passage information |

The second RWT subskill that focuses on passage evidence is **2d. Connecting Claims With Evidence**. Unlike **2c** questions, **2d** questions do not ask you to determine whether a passage claim is supported by evidence. Instead, these questions presume a claim is supported and ask you to identify that support. This process can also take place in the other direction, starting from the evidence and asking you to identify the supported claim. Hence, these sorts of questions might be phrased in any of the following ways:

- In the first paragraph the author asserts [x]. What is the evidence for this claim? (**claim to evidence**)
- The example of [y] is used as evidence for which of the following passage claims? (**evidence to claim**)
- The information about [z] most supports the passage claim that: (**evidence to claim**)

A less common variation of these questions relates to how a claim could be *tested*. For instance, you might run into a question like:

- Which of the following passage claims could most easily be challenged by contradictory evidence? (**claim to evidence**)

Although this question may seem out of place compared with the others, answering it depends on recognizing the type of evidence that claims are based on. Accordingly, that process still involves connecting passage claims with their supporting evidence.

# Method for Answering 2d. Connecting Claims With Evidence Questions

## Step 1: Identify the Relevant Claim in the Passage

With this type of question we are looking for a claim that we know is supported by additional passage information. Accordingly, the first step is to locate the claim and examine how it relates to other parts of the passage. For example, the claim could appear as one of several points that the author addresses in order, or introduce an overarching theme to be considered.

## Step 2: Review What Else Is Said About That Claim

The next step is to review what else is said about the claim in question. Often, this additional information may occur just before or after the original claim; if not, the way that claim is presented may suggest where the additional information can be found. Since some of this information is specifically meant to be supporting evidence, it will likely use wording similar to that of the original claim or constitute an example. Alternatively, it may be introduced by phrases that explicitly signal a return to the original claim.

## Step 3: Match the Claim to Its Specific Supporting Evidence

Finally, the connections you identified in Step 2 will reveal the specific passage information that serves as evidence for the claim. In the context of the passage, this information will provide a supporting reason for why that claim should be accepted. Only one of the answer choices will reflect this information.

As stated earlier, a question may instead go in the opposite direction by specifying the passage evidence and asking you to identify the claim this evidence supports. Aside from that difference, the method for answering these questions is the same.

## Subskill 2d. Connecting Claims With Evidence
## (3 Practice Questions)

We now look at three examples from the UWorld Qbank. In each case, the passage and question are first given without commentary, allowing you to practice applying the method yourself. Then, the passage and question are presented again, this time with annotations of the passage (see the CARS Passage Booklet) and a step-by-step explanation of the question using the described method.

## Passage A: The Knights Templar

The seal of the Knights Templar depicts two knights astride a single horse, a visual testament of the order's poverty at its inception in 1119. Nevertheless, these Knights of the Temple—who swore oaths not only of poverty but of chastity, loyalty, and bravery—would eventually become one of the wealthiest and most powerful organizations in the medieval world. So far-reaching was their strength and acclaim that their destruction must have seemed as sudden and surprising as it was utter and irrevocable. The signs of danger could not have been wholly invisible, however, as the Templars' growing influence became perceived as a threat to European rulers.

The first Templars were nine knights who took an oath to defend the Holy Land and any pilgrims who journeyed there after the First Crusade. Having secured a small benefice from Jerusalem's King Baldwin II, the knights inaugurated their mission at the site of the great Temple of Solomon. The order quickly attracted widespread admiration as well as many recruits from crusaders and other knights. Within a year of its founding, the order received a financial endowment from the deeply impressed Count of Anjou, whose example was soon followed by other nobles and monarchs. As early as 1128, the Templars even gained official papal recognition, and their wealth, holdings, and numbers swelled both in the Holy Land and throughout Europe, especially in France and England.

However, this growing power contained the seeds of the order's downfall. Although the Templars were generally held in high esteem, the passage of time saw censure and suspicion directed toward them. The failure of the disastrous siege of Ascalon in 1153 was attributed by some to Templar greed. Similarly, in 1208 Pope Innocent III condemned the wickedness he believed to exist within their ranks. Moreover, their increasingly elevated status brought them into conflict with established authorities. One revealing example occurred in 1252 when, because of the Templars' "many liberties" as well as their "pride and haughtiness," England's King Henry III proposed to curb the order's strength by reclaiming some of its possessions. The Templars' response was unambiguous: "So long as thou dost exercise justice thou wilt reign; but if thou infringe it, thou wilt cease to be King!"

Ultimately, the impoverished Philip the Fair of France joined forces with Pope Clement V to engineer the Templars' downfall beginning in 1307. Conspiring to seize the order's wealth, Clement invited the Templar Master, Jacques de Molay, to meet with him on the pretext of organizing a new crusade to retake the Holy Land. Shortly thereafter, Philip's forces arrested de Molay and his knights on charges the preponderance of which were almost certainly fabricated. Ranging from the mundane to the unspeakably perverse, the accusations even included an incredible entry citing "every crime and abomination that can be committed." Suffering tortures nearly as horrific as the acts of which they were accused, many Templars confessed.

Not everyone believed these charges. Despite Philip's urging, Edward II of England remained convinced that the accusations were false, a view seemingly shared by other rulers. Nevertheless, Clement ordered Edward to extract confessions, a task the king tried to carry out with some measure of mercy. By 1313, the Templar Order had been dissolved by papal decree, and many of its members were dead. The following year, Jacques de Molay was burned at the stake after declaring that the Templars' confessions were lies obtained under torture. Stories would spread that as he died, he condemned Clement and Philip to join him within a year. Although it is impossible to say whether he truly called divine vengeance down upon them, within a few months' time both pope and king had gone to their graves.

*The Knights Templar* ©UWorld

Annotations for this passage can be found in the CARS Passage Booklet.

## 2d. Connecting Claims With Evidence Question 1

Which of the following passage claims most strongly supports the assertion that the Templars' influence seemed threatening to European rulers?

A. Nobles throughout Europe provided money to the Order.
B. The Templars defied the attempted reforms of Henry III.
C. The Templars were absorbing crusaders and other knights into their ranks.
D. Clement V conspired to condemn the Templars on false charges.

See next page for the *strategy-based explanation* of this question.

## 2d. Connecting Claims With Evidence Question 1
## Strategy-based Explanation

Which of the following passage claims most strongly supports the assertion that the Templars' influence seemed threatening to European rulers?

  A. Nobles throughout Europe provided money to the Order.
  B. The Templars defied the attempted reforms of Henry III.
  C. The Templars were absorbing crusaders and other knights into their ranks.
  D. Clement V conspired to condemn the Templars on false charges.

The question provides a passage assertion and asks us to identify which other passage claim functions as its supporting evidence. By looking at the assertion the question asks about, we can come up with a reasonable idea of the type of claim that would support it. Specifically, we are looking for evidence that the Templars' influence seemed threatening to European rulers. Thus, the supporting claim would likely discuss either rulers' reactions to Templar activity or tension between the two groups.

## Applying the Method

*Passage Excerpt*

[P1] The seal of the Knights Templar depicts two knights astride a single horse, a visual testament of the order's poverty at its inception in 1119. Nevertheless, these Knights of the Temple—who swore oaths not only of poverty but of chastity, loyalty, and bravery—would eventually become one of the wealthiest and most powerful organizations in the medieval world. So far-reaching was their strength and acclaim that their destruction must have seemed as sudden and surprising as it was utter and irrevocable. The signs of danger could not have been wholly invisible, however, as the Templars' growing influence became perceived as a threat to European rulers.

[P3] However, **this growing power** contained the seeds of the order's downfall. Although the Templars were generally held in high esteem, the passage of time saw censure and suspicion directed toward them. The failure of the disastrous siege of Ascalon in 1153 was attributed by some to Templar greed. Similarly, in 1208 Pope Innocent III condemned the wickedness he believed to exist within their ranks. Moreover, **their increasingly elevated status brought them into conflict with established authorities**. **One revealing example** occurred in

### Step 1: Identify the Relevant Claim in the Passage

The claim that the question references appears at the end of Paragraph 1: the Templars' growing influence became perceived as a threat to European rulers. Although the introduction seems to build up to this claim, no supporting evidence is given for it there. Thus, it is likely a theme that the author will revisit later in the passage.

### Step 2: Review What Else Is Said About That Claim

We can recognize where the author revisits the original claim by the similar language used in Paragraph 3: the Templars' **growing power** and **elevated status brought them into conflict with established authorities**.

1252 when, because of the Templars' "many liberties" as well as their "pride and haughtiness," **England's King Henry III proposed to curb the order's strength** by reclaiming some of its possessions. **The Templars' response was unambiguous:** "So long as thou dost exercise justice thou wilt reign; but if thou infringe it, thou wilt cease to be King!"

> The author then illustrates the effects of this conflict by emphasizing **one revealing example**. When **King Henry III attempted to rein in the Templars' power, the Templars responded with a threat.**

### Step 3: Match the Claim to Its Specific Supporting Evidence

We have now identified both the original claim and where the author elaborates on it in the passage. The assertion that the Templars' influence seemed threatening to European rulers is strongly supported by the claim that the Templars threatened to remove King Henry III from power when he tried to oppose them. Accordingly, the correct answer is **Choice B**, *The Templars defied the attempted reforms of Henry III*.

There is another answer choice that mentions conflict with a ruler: Choice D refers to how *Clement V conspired to condemn the Templars on false charges*. However, the passage never depicts the Templars as threatening Clement, while it does show them threatening to depose Henry III. Consequently, **Choice B** is the better answer.

For an alternative method of explanation based more specifically on passage evidence, you can view this question in the UWorld Qbank (sold separately).

# Passage L: When Defense Is Indefensible

Suppose a prosecutor is considering whether to bring a case to trial. He is not sure that the suspect is guilty—in fact, based on the evidence, it's more likely that the suspect is *not* guilty. Nevertheless, he feels confident he can secure a guilty verdict. His powers of persuasion are considerable, and there's a good chance he could trick a jury into believing the evidence is strong instead of weak. In addition, the case is high profile and could be very lucrative; winning would likely lead to a substantial raise or promotion. He decides to charge the suspect, and ultimately succeeds in persuading the jury to convict.

Looking at this situation, most of us would easily judge the prosecutor as extremely unethical. His conduct is outrageous and wrong—he clearly acted with corrupt intent, perpetrating injustice in order to profit financially. Why is it not shocking, then, that we tolerate the mirror image of this behavior from defense attorneys? For they engage in the same outrageous conduct, only on the other side. Paid handsomely to represent even the vilest of clients, they apply their oratorical prowess to manipulating jury perception, keeping the guilty free and unpunished in exchange for money and status. To the extent that this behavior takes place, are some defense attorneys as unethical as our hypothetical prosecutor?

It is worth distinguishing two senses of the word "ethical" here. For there are standards of *professional* ethics to which any attorney must conform, including standards particular to the defense. Most relevant to our purposes, a lawyer is obligated to provide their client with a "zealous defense." In other words, once an attorney takes on a client, they are ethically bound to promote that client's rights, interests, or innocence—in fact, *not* to do so would be *unethical*. Thus, one might try to suggest that this obligation undermines the claim that some defense attorneys act unethically.

However, meeting that professional standard is not the same as being ethical in the general sense of the word. The standard depends on the condition: *once an attorney takes on a client*. With the exception of court-appointed attorneys or public defenders, who are assigned to provide representation to those who would otherwise lack it, an attorney is never required to represent a defendant. Therefore, meeting one's obligations as a defense attorney does not necessarily make one ethical, because the choice to accept a specific case (and thus to incur those obligations in the first place) may itself be an unethical act.

Moreover, the role of court-appointed attorneys is to help protect the rights of citizens who cannot secure their own representation, usually for financial reasons. Although preserving those rights is necessary to uphold justice, this situation highlights how wealth and class contribute to *injustice*. While some defendants possess the means to hire top-level private lawyers, others must depend on public servants—frequently less experienced lawyers from overloaded, understaffed agencies. As a result, the rich are more likely to escape conviction even when they are guilty, and the poor are more likely to be convicted even when they are innocent.

It is doubtful that private defense attorneys could be somehow forbidden from representing guilty clients. Hence, the needed reforms to the system must come from individual attorneys committing to work for the right reasons. For those who strive to ensure citizens' rights, or who truly believe their clients are innocent, providing a defense is a noble undertaking. But for those whose overriding motivation is greed, that legally "zealous defense" is ethically indefensible.

*When Defense Is Indefensible* ©UWorld

Annotations for this passage can be found in the CARS Passage Booklet.

## 2d. Connecting Claims With Evidence Question 2

The claim that most people would easily view the hypothetical prosecutor as unethical is used to support the assertion that:

- A. defense attorneys represent even the vilest of clients.
- B. prosecutors can be just as unethical as defense attorneys.
- C. defense attorneys are mirror images of the hypothetical prosecutor.
- D. it is strange that we accept the behavior of defense attorneys.

See next page for the *strategy-based explanation* of this question.

> ## 2d. Connecting Claims With Evidence Question 2
> ## Strategy-based Explanation
>
> The claim that most people would easily view the hypothetical prosecutor as unethical is used to support the assertion that:
>
> A. defense attorneys represent even the vilest of clients.
> B. prosecutors can be just as unethical as defense attorneys.
> C. defense attorneys are mirror images of the hypothetical prosecutor.
> D. it is strange that we accept the behavior of defense attorneys.

This question gives us the supporting claim first, then asks us to connect it to the claim it is evidence for. Since describing the hypothetical prosecutor takes up the entire introductory paragraph, you likely remember that the claim being asked about comes at the beginning of Paragraph 2. Thus, answering the question is simply a matter of reviewing that paragraph to determine what assertion the claim helps establish.

## Applying the Method

*Passage Excerpt*

[P2] **Looking at this situation, most of us would easily judge the prosecutor as extremely unethical.** His conduct is outrageous and wrong—he clearly acted with corrupt intent, perpetrating injustice in order to profit financially. **Why is it not shocking, then, that we tolerate the mirror image of this behavior from defense attorneys?** For they engage in the same outrageous conduct, only on the other side. Paid handsomely to represent even the vilest of clients, they apply their oratorical prowess to manipulating jury perception, keeping the guilty free and unpunished in exchange for money and status. To the extent that this behavior takes place, are some defense attorneys as unethical as our hypothetical prosecutor?

### Step 1: Identify the Relevant Claim in the Passage

The claim in question can be found at the beginning of Paragraph 2: **Looking at this situation, most of us would easily judge the prosecutor as extremely unethical.**

The author then elaborates on this judgment of the prosecutor, asserting: His conduct is outrageous and wrong—he clearly acted with corrupt intent, perpetrating injustice in order to profit financially.

Thus, the author makes a claim about how most people would view this conduct, then explains why the conduct is so wrong.

### Step 2: Review What Else Is Said About That Claim

Immediately after describing the prosecutor's corrupt conduct, the author poses a question: **Why is it not shocking, then, that we tolerate the mirror image of this behavior from defense attorneys?**

As before, the author then elaborates: For they engage in the same outrageous conduct, only on the other side. Paid handsomely to represent even the vilest of clients…in exchange for money and status.

Accordingly, the author uses a structure similar to that of the previous sentences: asserting how the described conduct should be judged, then further explaining why it is wrong.

## Step 3: Match the Claim to Its Specific Supporting Evidence

With this question, the connection between the evidence and the claim it supports is straightforward. The author asserts that **most of us would easily judge the prosecutor as extremely unethical**, then almost immediately asks: **Why is it not shocking, then, that we tolerate the mirror image of this behavior from defense attorneys?** The surrounding sentences reinforce this connection by describing how each party's conduct is similarly corrupt: prosecuting an innocent person in order to profit financially, and defending a guilty person in exchange for money and status.

Accordingly, the author is pointing out what they see as a discrepancy in people's attitudes. If we view the hypothetical prosecutor as unethical, then we should also view some defense attorneys as unethical, but we don't. Therefore, the claim that most people would easily view the hypothetical prosecutor as unethical is used to support **Choice D**, *it is strange that we accept the behavior of defense attorneys*.

### Consistent Versus Correct

You probably noticed that all four answer choices in this question are stated or implied in the passage. However, only one is specifically supported by the claim being asked about. For example, one of the wrong answers is Choice C, *defense attorneys are mirror images of the hypothetical prosecutor*. This answer implies that the author makes the following argument:

> 'We believe that the prosecutor acts unethically. So, we should also believe that some defense attorneys act like his mirror image.'

However, examining the passage shows that the author actually argues as follows:

> 'We believe that the prosecutor acts unethically. Some defense attorneys act like his mirror image. So, we should also believe that **those defense attorneys act unethically**.'

Accordingly, although Choice C is *consistent* with passage information, it is nevertheless incorrect. You will find that this is the case with many wrong answers in CARS. Thus, it is important to make sure that when you choose an answer choice, it doesn't just align with the passage—it must truly *answer the question*.

For an alternative method of explanation based more specifically on passage evidence, you can view this question in the UWorld Qbank (sold separately).

## Passage D: American Local Motives

Locomotives were invented in England, with the first major railroad connecting Liverpool and Manchester in 1830. However, it was in America that railroads would be put to the greatest use in the nineteenth century. On May 10, 1869, the Union Pacific and Central Pacific lines met at Promontory Point, Utah, joining from opposite directions to complete a years-long project—the Transcontinental Railroad. This momentous event connected the eastern half of the United States with its western frontier and facilitated the construction of additional lines in between. As a result, journeys that had previously taken several months by horse and carriage now required less than a week's travel. By 1887 there were nearly 164,000 miles of railroad tracks in America, and by 1916 that number had swelled to over 254,000.

While the United States still has the largest railroad network in the world, it operates largely in the background of American life, and citizens no longer view trains with the sense of importance those machines once commanded. Nevertheless, the economic and industrial advantages those citizens enjoy today would not have been possible without America's history of trains; as Tom Zoellner reminds us, "Under the skin of modernity lies a skeleton of railroad tracks." Although airplanes and automobiles have now assumed greater prominence, the time has arrived for the resurgence of railroads. A revitalized and advanced railway system would confer numerous essential benefits on both the United States and the globe.

The chief obstacles to garnering support for such a project are the current dominance of the automobile and the languishing technology of existing railroads. In a sense these two obstacles are one, as American dependence on personal automobiles is partially due to the paucity of rapid public transportation. The railroads of Europe and Japan, by comparison, have vastly outpaced their American counterparts. Japan has operated high-speed rail lines continuously since 1964, and in 2007, a French train set a record of 357 miles per hour. While that speed was achieved under tightly controlled conditions, it still speaks to the great disparity in railroad development between the United States and other countries since the mid-twentieth century. British trains travel at speeds much higher than those in America, where both the trains themselves and the infrastructure to support them have simply been allowed to fall behind. In much of Europe it is common for trains to travel at close to 200 miles per hour.

To invest in a modern network of railroads would improve the United States in much the same way that the first railroads did in the nineteenth and early twentieth centuries. A high-speed passenger rail system would dramatically transform American life as travel between cities and states became quicker and more convenient, encouraging commerce, business, and tourism. Such a system would also make important strides in environmental preservation. According to a 2007 British study, "CO2 emissions from aircraft operations are...at least five times greater" than those from high-speed trains. For similar reasons, Osaka, Japan, was ranked as "the best…green transportation city in Asia" by the 2011 *Green City Index*. As Lee-in Chen Chiu notes in *The Kyoto Economic Review*, Osakans travel by railway more than twice as much as they travel by car.

It is true that developing a countrywide high-speed rail system would come with significant costs. However, that was also true of the original Transcontinental Railroad, as indeed it is with virtually any great project undertaken for the public good. We should thus move ahead with confidence that the rewards will outweigh the expenditure as citizens increasingly choose to travel by train. Both for society's gain and the crucial well-being of the planet, our path forward should proceed upon rails.

*American Local Motives* ©UWorld

Annotations for this passage can be found in the CARS Passage Booklet.

## 2d. Connecting Claims With Evidence Question 3
### Strategy-based Explanation

Of the following passage claims, which could most easily be confirmed or refuted?

- A. Americans no longer view trains with the sense of importance they once commanded.
- B. The railroads of Europe and Japan have vastly outpaced their American counterparts.
- C. An advanced railway system would benefit America economically.
- D. A high-speed passenger rail system would dramatically transform American life.

See next page for the *strategy-based explanation* of this question.

# 2d. Connecting Claims With Evidence Question 3
## Strategy-based Explanation

Of the following passage claims, which could most easily be confirmed or refuted?

A. Americans no longer view trains with the sense of importance they once commanded.
B. The railroads of Europe and Japan have vastly outpaced their American counterparts.
C. An advanced railway system would benefit America economically.
D. A high-speed passenger rail system would dramatically transform American life.

As mentioned in the introduction to this subskill, sometimes a question asks about the possibility of testing passage claims. In this example, we are asked which of the given claims would be easiest to confirm or refute, or which could most directly be verified. To answer that question, we need to review each answer choice to determine what type of evidence could be used to support it. Note that it is irrelevant whether such evidence actually appears in the passage; the issue is only the degree to which each claim could potentially be proven or disproven.

## Applying the Method

*Passage Excerpt*

**[P2]** While the United States still has the largest railroad network in the world, it operates largely in the background of American life, and **citizens no longer view trains with the sense of importance those machines once commanded**. Nevertheless, the economic and industrial advantages those citizens enjoy today would not have been possible without America's history of trains; as Tom Zoellner reminds us, "Under the skin of modernity lies a skeleton of railroad tracks."

**[P3]** The chief obstacles to garnering support for such a project are the current dominance of the automobile and the languishing technology of existing railroads. In a sense these two obstacles are one, as American dependence on personal automobiles is partially due to the paucity of rapid public transportation. **The railroads of Europe and Japan, by comparison, have vastly outpaced their American counterparts**. Japan has operated high-speed rail lines continuously since 1964, and in 2007, a French train set a record of 357 miles per hour. While that speed was achieved under tightly controlled conditions, it still speaks to the great disparity in railroad development between the United States and other countries since the mid-twentieth century.

### Step 1: Identify the Relevant Claim in the Passage

### Step 2: Review What Else Is (or Could Be) Said About That Claim

**Choice A**

"Americans no longer view trains with the sense of importance they once commanded."

- Paragraph 2 contains the claim that citizens no longer view trains with the sense of importance those machines once commanded.

**What Else Could Be Said About That Claim**

- Paragraph 3 refers to the American dependence on personal automobiles. This dependence would be consistent with a view that trains are less important than they once were. However, this fact does not in itself tell us whether people hold that view; it only describes their behavior.

- Accordingly, the evidence of such a view would either have to be inferred from behavior in this way, or concluded from a study that asked about people's attitudes toward trains.

**Choice B**

"The railroads of Europe and Japan have vastly outpaced their American counterparts."

- According to Paragraph 3, **The railroads of Europe and Japan, by comparison, have**

British trains travel at speeds much higher than those in America, where both the trains themselves and the infrastructure to support them have simply been allowed to fall behind. In much of Europe it is common for trains to travel at close to 200 miles per hour.

[P4] To invest in a modern network of railroads would improve the United States in much the same way that the first railroads did in the nineteenth and early twentieth centuries. **A high-speed passenger rail system would dramatically transform American life as travel between cities and states became quicker and more convenient**, **encouraging commerce, business, and tourism**. Such a system would also make important strides in environmental preservation.

**vastly outpaced their American counterparts.**

**What Else Could Be Said About That Claim**

- To illustrate this comparison, the passage lists data about the length of time Japan has operated high-speed trains, and the speeds at which French and British trains travel.

### Choice D

"A high-speed passenger rail system would dramatically transform American life."

- Paragraph 4 states that **a high-speed passenger rail system would dramatically transform American life as travel between cities and states became quicker and more convenient**.

**What Else Could Be Said About That Claim**

- The passage does not provide details about the extent of this claimed transformation. Such details would presumably produce measurable data *after* a high-speed rail system had been built, but currently the claim is a projection about the future.

### Choice C

"An advanced railway system would benefit America economically."

- Paragraph 4 suggests that high-speed rail would lead to economic gains by **encouraging commerce, business, and tourism**.

**What Else Could Be Said About That Claim**

- As with Choice D, such economic effects would generate measurable statistics in the future. However, any current data could only describe what is assumed or predicted to result.

### Step 3: Match the Claim to Its Specific Supporting Evidence

In reviewing each claim, we found that only one is based on currently existing, easily measurable data: **Choice B**, *The railroads of Europe and Japan have vastly outpaced their American counterparts*. Facts about the speed and operation of those railroads could be directly compared with similar data about railroads in America, allowing us to empirically verify any comparative claims about them. Accordingly, this is the passage claim that could most easily be confirmed or refuted.

For an alternative method of explanation based more specifically on passage evidence, you can view this question in the UWorld Qbank (sold separately).

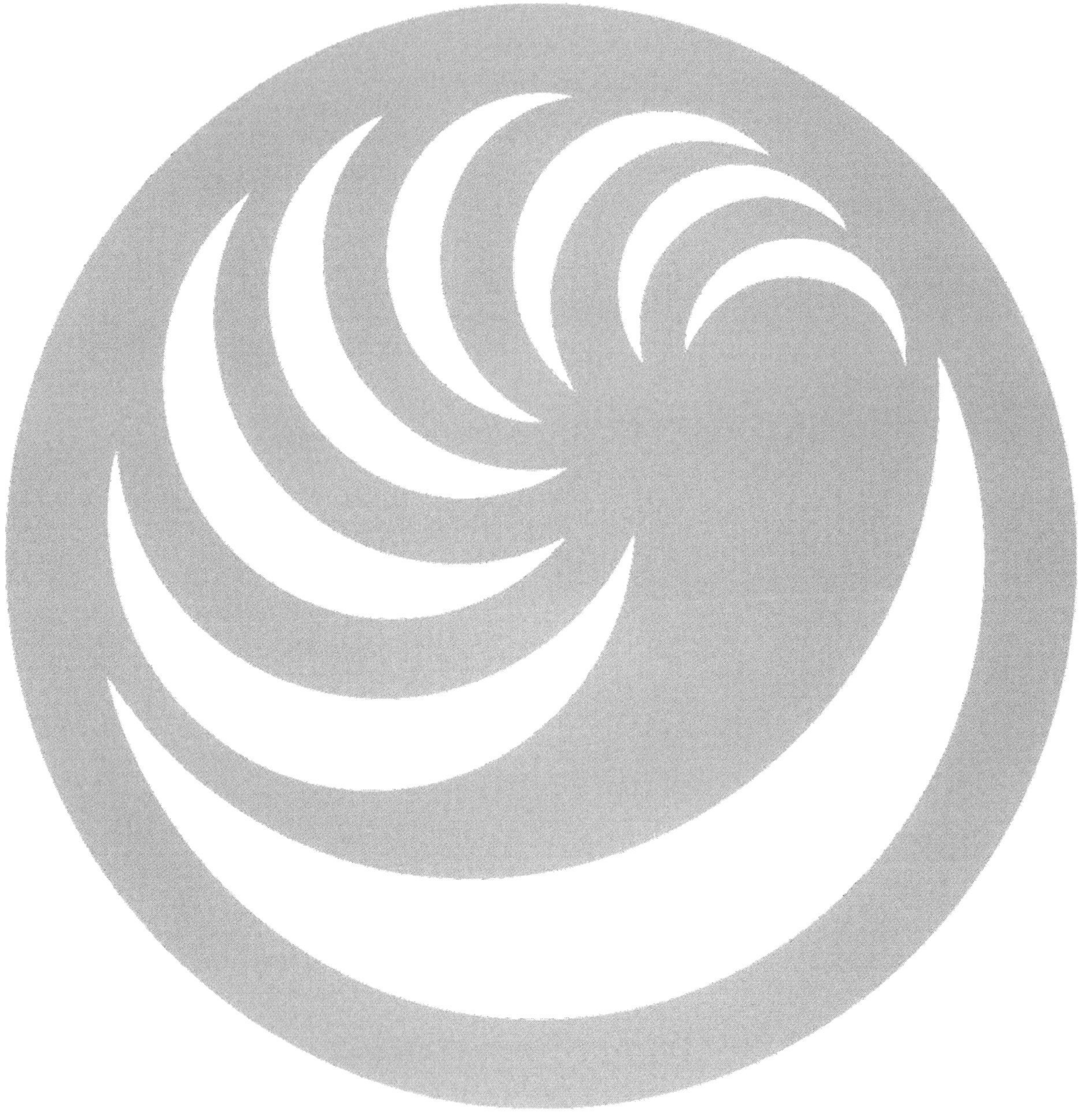

Lesson 6.5

# RWT Subskill 2e. Determining Passage Perspectives

| Skill 2: Reasoning Within the Text ||
|---|---|
| **Subskill** | **Student Objective** |
| 2a. Logical Relationships Within Passage | Identify logical relationships between passage claims |
| 2b. Function of Passage Claim | Determine what function a particular claim serves in the passage |
| 2c. Extent of Passage Evidence | Determine the extent to which the passage provides evidence for its claims |
| 2d. Connecting Claims With Evidence | Connect passage claims with supporting passage evidence |
| **2e. Determining Passage Perspectives** | **Determine the perspective of the author or another source in the passage** |
| 2f. Drawing Additional Inferences | Draw additional inferences based on passage information |

To answer **2e. Determining Passage Perspectives** questions, you must figure out what the author or another source in the passage believes or feels about a particular topic. Some examples of such questions are:

- Based on the passage, which of the following most accurately represents the author's view regarding the Frank Marshall chess game of 1912?
- Based on the information in Paragraph 5, with which of the following statements would the passage author most likely agree?
- Which of the following best describes how the passage author treats the idea that the differences between the stories of *Zadig* and *Candide* might reflect changes in Voltaire's own outlook over time?

As you can tell, these kinds of questions are similar to those you encountered in **1e. Identifying Passage Perspectives** (see Lesson 5.5). However, answering **2e** questions may involve more complex reasoning or a larger amount of passage information, which is why it constitutes an example of Reasoning Within the Text.

## The Difference Between 1e and 2e Questions

As a way to conceptualize the difference between **1e** and **2e** questions, imagine two passages that express a view about a particular artist. In the first passage, the author states that 'the artist's work has hidden depths.' In the second passage, the author states: 'Much has been written about the artist's work. Yet these observations have typically been limited to the surface level meaning.'

You can probably see that the author's view of the artist's work is the same in both cases. However, recognizing this view takes more mental effort in the second example. So, while either passage might include a question about the author's perspective, the first would require you only to comprehend the passage, but the second would also require you to reason about the author's claim. In other words, the first passage would generate a **CMP 1e. Identifying Passage Perspectives** question, while the second passage would generate a **RWT 2e. Determining Passage Perspectives** question.

Nevertheless, the method for answering either version of these questions is essentially the same—some of the questions may just involve extra reasoning.

## Method for Answering 2e. Determining Passage Perspectives Questions

### Step 1: Take Note of Viewpoint Indicators

As with **1e** questions, we can determine a source's position by taking note of **Viewpoint Indicators**: words and phrases that reveal perspectives or introduce information that reveals them. At times, the indicators for **2e** questions may be less obvious than those for **1e** questions. They may also be spread over multiple parts of the passage instead of being concentrated in one. Regardless, we are still looking for the same types of words and phrases, those that:

- **evaluate an idea**

    "However, the connotations attached to these uses are largely unjustified."

    "The most important point to recognize in this regard is that…"

- **refer to other views**

    "Although some deem Sun Bin's thinking to lack the philosophical depth displayed by his predecessor, the two strategists in fact bear many similarities."

- **signal a conclusion**

    "…those emphases clearly reflect the intended meaning."

    "Hence, the profit obtained from exporting these and other goods could plausibly account for…"

You may find it useful to highlight these indicators when you come across them in the passage so you can more easily locate the views when needed.

**Step 2: Determine the Source's Belief or Attitude**

By taking note of the Viewpoint Indicators in the passage, you will see how the author or other source discusses and characterizes the topic being asked about, thereby conveying the perspective you are looking for. When you review the answer choices, you will find that only one of them reflects this same perspective.

## Subskill 2e. Determining Passage Perspectives
## (5 Practice Questions)

We now look at five examples from the UWorld Qbank. In each case, the passage and question are first given without commentary, allowing you to practice applying the method yourself. Then, the passage and question are presented again, this time with annotations of the passage (see the CARS Passage Booklet) and a step-by-step explanation of the question using the described method.

## Passage A: The Knights Templar

The seal of the Knights Templar depicts two knights astride a single horse, a visual testament of the order's poverty at its inception in 1119. Nevertheless, these Knights of the Temple—who swore oaths not only of poverty but of chastity, loyalty, and bravery—would eventually become one of the wealthiest and most powerful organizations in the medieval world. So far-reaching was their strength and acclaim that their destruction must have seemed as sudden and surprising as it was utter and irrevocable. The signs of danger could not have been wholly invisible, however, as the Templars' growing influence became perceived as a threat to European rulers.

The first Templars were nine knights who took an oath to defend the Holy Land and any pilgrims who journeyed there after the First Crusade. Having secured a small benefice from Jerusalem's King Baldwin II, the knights inaugurated their mission at the site of the great Temple of Solomon. The order quickly attracted widespread admiration as well as many recruits from crusaders and other knights. Within a year of its founding, the order received a financial endowment from the deeply impressed Count of Anjou, whose example was soon followed by other nobles and monarchs. As early as 1128, the Templars even gained official papal recognition, and their wealth, holdings, and numbers swelled both in the Holy Land and throughout Europe, especially in France and England.

However, this growing power contained the seeds of the order's downfall. Although the Templars were generally held in high esteem, the passage of time saw censure and suspicion directed toward them. The failure of the disastrous siege of Ascalon in 1153 was attributed by some to Templar greed. Similarly, in 1208 Pope Innocent III condemned the wickedness he believed to exist within their ranks. Moreover, their increasingly elevated status brought them into conflict with established authorities. One revealing example occurred in 1252 when, because of the Templars' "many liberties" as well as their "pride and haughtiness," England's King Henry III proposed to curb the order's strength by reclaiming some of its possessions. The Templars' response was unambiguous: "So long as thou dost exercise justice thou wilt reign; but if thou infringe it, thou wilt cease to be King!"

Ultimately, the impoverished Philip the Fair of France joined forces with Pope Clement V to engineer the Templars' downfall beginning in 1307. Conspiring to seize the order's wealth, Clement invited the Templar Master, Jacques de Molay, to meet with him on the pretext of organizing a new crusade to retake the Holy Land. Shortly thereafter, Philip's forces arrested de Molay and his knights on charges the preponderance of which were almost certainly fabricated. Ranging from the mundane to the unspeakably perverse, the accusations even included an incredible entry citing "every crime and abomination that can be committed." Suffering tortures nearly as horrific as the acts of which they were accused, many Templars confessed.

Not everyone believed these charges. Despite Philip's urging, Edward II of England remained convinced that the accusations were false, a view seemingly shared by other rulers. Nevertheless, Clement ordered Edward to extract confessions, a task the king tried to carry out with some measure of mercy. By 1313, the Templar Order had been dissolved by papal decree, and many of its members were dead. The following year, Jacques de Molay was burned at the stake after declaring that the Templars' confessions were lies obtained under torture. Stories would spread that as he died, he condemned Clement and Philip to join him within a year. Although it is impossible to say whether he truly called divine vengeance down upon them, within a few months' time both pope and king had gone to their graves.

*The Knights Templar* ©UWorld

Annotations for this passage can be found in the CARS Passage Booklet.

Chapter 6: Skill 2 – Reasoning Within the Text

## 2e. Determining Passage Perspectives Question 1

Which of the following items of passage information most clearly indicates that the author believes the Templars were innocent of the accusations against them?

- A. The Templars swore oaths of poverty, chastity, loyalty, and bravery.
- B. The list of charges included every crime and abomination.
- C. Jacques de Molay declared that the Templars' confessions were lies.
- D. The rulers of many countries did not believe the accusations.

See next page for the *strategy-based explanation* of this question.

Chapter 6: Skill 2 – Reasoning Within the Text

> ### 2e. Determining Passage Perspectives Question 1
> ### Strategy-based Explanation
>
> Which of the following items of passage information most clearly indicates that the author believes the Templars were innocent of the accusations against them?
>
> A. The Templars swore oaths of poverty, chastity, loyalty, and bravery.
> B. The list of charges included every crime and abomination.
> C. Jacques de Molay declared that the Templars' confessions were lies.
> D. The rulers of many countries did not believe the accusations.

Phrased in simpler terms, the question asks which passage claim best reveals the author's view that the Templars were innocent. Accordingly, while each answer choice could represent a reason to believe the Templars were innocent, only one will represent the *author's* reason for holding that view.

## Applying the Method

*Passage Excerpt*

**[P1]** The seal of the Knights Templar depicts two knights astride a single horse, a visual testament of the order's poverty at its inception in 1119. Nevertheless, **these Knights of the Temple—who swore oaths not only of poverty but of chastity, loyalty, and bravery**—would eventually become one of the wealthiest and most powerful organizations in the medieval world. So far-reaching was their strength and acclaim that their destruction must have seemed as sudden and surprising as it was utter and irrevocable. The signs of danger could not have been wholly invisible, however, as the Templars' growing influence became perceived as a threat to European rulers.

**[P4]** Ultimately, the impoverished Philip the Fair of France joined forces with Pope Clement V to engineer the Templars' downfall beginning in 1307. Conspiring to seize the order's wealth, Clement invited the Templar Master, Jacques de Molay, to meet with him on the pretext of organizing a new crusade to retake the Holy Land. Shortly thereafter, Philip's forces arrested de Molay and his knights on charges the preponderance of which were almost certainly fabricated. Ranging from the mundane

### Step 1: Take Note of Viewpoint Indicators

While all of the **answer choices** do appear in the passage, only one of them is accompanied by viewpoint indicators that specifically show the author's perspective.

**Viewpoint Indicator**

In Paragraph 4, the author asserts that the charges against the Templars **were almost certainly fabricated**, conveying the author's view that these charges were probably false.

| | |
|---|---|
| to the unspeakably perverse, **the accusations even included an incredible entry** citing "**every crime and abomination that can be committed.**" Suffering tortures nearly as horrific as the acts of which they were accused, many Templars confessed. | **Viewpoint Indicator**<br><br>The author then elaborates on their view with an example, stating that the charges **even included an incredible entry** that accused the Templars of "**every crime and abomination that can be committed.**" |
| **[P5]** Not everyone believed these charges. Despite Philip's urging, **Edward II of England remained convinced that the accusations were false, a view seemingly shared by other rulers**. Nevertheless, Clement ordered Edward to extract confessions, a task the king tried to carry out with some measure of mercy. By 1313, the Templar Order had been dissolved by papal decree, and many of its members were dead. The following year, **Jacques de Molay was burned at the stake after declaring that the Templars' confessions were lies obtained under torture**. Stories would spread that as he died, he condemned Clement and Philip to join him within a year. Although it is impossible to say whether he truly called divine vengeance down upon them, within a few months' time both pope and king had gone to their graves. | |

### Step 2: Determine the Source's Belief or Attitude

By noting the author's viewpoint indicators, we can determine why they believe the Templars were innocent. The author states that the charges against the Templars **were almost certainly fabricated** or made up, then highlights how some of those charges were **incredible** or beyond belief. Accordingly, the passage information that most clearly indicates that the author believes the Templars were innocent is **Choice B**, *the list of charges included every crime and abomination*.

You probably noticed that there are also viewpoint indicators corresponding to Choice D. Paragraph 5 tells us that Edward II "remained convinced" that the accusations against the Templars were false, and that this was "a view seemingly shared by other rulers." However, while this statement indicates what *those rulers* believed, it is not a viewpoint indicator *for the author*. Rather, the author reports those rulers' shared belief but does not express agreement or disagreement with it. By contrast, the author's own view is indicated in connection with **Choice B**, which is therefore the correct answer.

For an alternative method of explanation based more specifically on passage evidence, you can view this question in the UWorld Qbank (sold separately).

## Passage J: The Inkblots

For almost 100 years now, the psychological evaluation known as the Rorschach Inkblot Test has engendered much controversy, including skepticism about its value, questions about its scoring, and, especially, criticism of its interpretive methods as too subjective. Thus, the Rorschach test, which emerged from the same early twentieth-century zeitgeist that produced Einstein's physics, Freudian psychoanalysis, and abstract art, seems one of modernity's most misbegotten children. Destined never to be completely accepted or discredited, the test remains a perennial outlier in its field. Nevertheless, the inkblots' mystery and aesthetic appeal have caused them to be indelibly printed on our cultural fabric.

The now iconic inkblots were introduced to the world by Swiss psychiatrist Hermann Rorschach in his 1921 book *Psychodiagnostics*. As both director of the Herisau Asylum in Switzerland and an amateur artist, Rorschach was uniquely positioned to wed the new practice of psychoanalysis to the budding phenomenon of abstract art. For instance, reading Freud's work on dream symbolism prompted him to recall his childhood passion for a game based on inkblot art called *Klecksographie*. He was also cognizant that in a recently published dissertation, his colleague Szymon Hens had used inkblots to try to probe the imagination of research subjects; moreover, a few years earlier, the French psychologist and father of intelligence testing Alfred Binet had used them to measure creativity.

Motivated by these developments, the Herisau director decided to revisit that childhood pastime that had awakened his curiosity about how visual information is processed. In particular, he wondered why different people saw different things in the same image. Traditionally, psychoanalysts had relied on language for insights; however, as biographer Damion Searls reports, Rorschach's theories would exemplify the principle that "who we are is a matter less of what we say than of what we see." Indeed, through a process of perception termed pareidolia, the mind projects meaning onto images, detecting in them familiar objects or shapes. Consequently, what a person sees in an image reveals more about that person than about the image itself.

Rorschach experimented with countless inkblots, eventually selecting ten—five black on white, two also featuring some red, and three pastel-colored—to use with research subjects. For these perfectly symmetrical images—each of which he was said to have "meticulously designed to be as ambiguous and 'conflicted' as possible"—the primary question was always "What do *you* see?" Rorschach was especially careful to note how much attention individuals paid to various components of each inkblot (such as form, color, and a sense of movement) and whether they concentrated on details or the whole image. Having observed that his patients with schizophrenia gave distinctly different responses from the control group, Rorschach envisioned his experiment as a diagnostic tool for the disease. Nevertheless, he resisted the notion that its results could be used to assess personality. In fact, until his untimely death from a ruptured appendix in 1922, Rorschach referred to his project as an "interpretive form experiment" rather than a test. Ironically, however, by the 1960s, the Rorschach Inkblot Test was known chiefly as a personality assessment and had become the most frequently administered projective personality test in the US.

Rorschach's test has survived nearly incessant scrutiny, including a 2013 comprehensive study of all Rorschach test data and repeated revisions to its scoring, yet doubts about its validity and reliability persist. Much like the inkblots themselves—which tantalize us with the possibility of divulging the secrets of who we are and how we see the world—the test has (for better or worse) defied attempts to fix its meaning. Thus, what has been called "the twentieth century's most visionary synthesis of art and science" stands tempered by harsh criticism.

*The Inkblots* ©UWorld

Annotations for this passage can be found in the CARS Passage Booklet.

## 2e. Determining Passage Perspectives Question 2

The author's attitude toward the passage topic can best be described as:

A. dismissive.
B. critical.
C. affirming.
D. open.

See next page for the *strategy-based explanation* of this question.

> ## 2e. Determining Passage Perspectives Question 2
> ## Strategy-based Explanation
>
> The author's attitude toward the passage topic can best be described as:
>
> A. dismissive.
> B. critical.
> C. affirming.
> D. open.

This question asks us to determine the attitude or perspective the author has toward their topic, namely the Rorschach Inkblot Test. Thus, although the passage is primarily a descriptive account of the test and its development, we can expect to see ways in which the author's own thoughts or feelings are revealed.

## Applying the Method

*Passage Excerpt*

[P1] For almost 100 years now, the psychological evaluation known as the Rorschach Inkblot Test **has engendered much controversy**, including skepticism about its value, questions about its scoring, and, especially, criticism of its interpretive methods as too subjective. Thus, the Rorschach test, which emerged from the same early twentieth-century zeitgeist that produced Einstein's physics, Freudian psychoanalysis, and abstract art, seems one of modernity's most misbegotten children. **Destined never to be completely accepted or discredited**, the test remains **a perennial outlier in its field**. **Nevertheless**, the inkblots' **mystery and aesthetic appeal** have caused them to be **indelibly printed on our cultural fabric**.

[P5] Rorschach's test has **survived nearly incessant scrutiny**, including a 2013 comprehensive study of all Rorschach test data and repeated revisions to its scoring, yet **doubts about its validity and reliability persist**. Much like the inkblots themselves—which **tantalize us** with the possibility of divulging the secrets of who we are and how we see the world—the test has **(for better or worse)** defied attempts to fix its meaning. Thus, what has been called **"the twentieth century's most**

### Step 1: Take Note of Viewpoint Indicators

**Viewpoint Indicators**

The passage begins by describing areas of **controversy** surrounding the Rorschach Inkblot Test, noting it was **destined never to be completely accepted or discredited** and has remained **a perennial outlier in its field**. Thus, the author conveys various negative points about the test.

However, the author then mentions the inkblots' **mystery and aesthetic appeal,** asserting that they have become **indelibly printed on our cultural fabric.** In this way, the inkblots are also portrayed positively.

**Viewpoint Indicators**

The passage ends in much the same way it began. According to Paragraph 5, the Rorschach test **has survived nearly incessant scrutiny** though **doubts about its validity and reliability persist**. The inkblots themselves **tantalize us…for better or worse.**

Similarly, the test has been described as **"the twentieth century's most visionary synthesis of art and science,"** but it has also been **tempered by harsh criticism**.

**visionary synthesis of art and science" stands tempered by harsh criticism.** | Thus, the author again presents both positive and negative viewpoints of the test.

### Step 2: Determine the Source's Belief or Attitude

By taking note of the viewpoint indicators in the passage, we can see that the author consistently offers both positive and negative descriptions of the Rorschach test. For instance, the author notes that the test has **engendered much controversy** but also mentions how the inkblots **are indelibly printed on our cultural fabric**. Similarly, while the test has been subjected to **incessant scrutiny**, it has also been called **"the twentieth century's most visionary synthesis of art and science."**

Accordingly, the author does not take a stance on one side or the other but presents a balanced or mixed perspective toward the Rorschach Inkblot Test. Therefore, **Choice D,** *open*, would best summarize the author's attitude toward the passage topic.

For an alternative method of explanation based more specifically on passage evidence, you can view this question in the UWorld Qbank (sold separately).

# Passage E: The Divine Sign of Socrates

From his bare feet to his bald pate, the potentially shapeshifting figure of Socrates found in the literary tradition that arose after his controversial trial and death presents an intriguing array of oddities and unorthodoxies. Most conspicuously, his unshod and shabby sartorial state flaunted poverty at a time when the city of Athens had become obsessed with wealth and its trappings. Yet the philosopher's peculiar appearance was but a hint of the strange new calling he embraced. Inspired perhaps by the famous Delphic dictum "Know thyself," he embarked on a mission devoted to finding truth through dialogue. In what struck some as a dangerous new method of inquiry, he subjected nearly everyone he encountered to intense cross-examination, mercilessly exposing the ignorance of his interlocutors. Moreover, in a culture that still put stock in magic, the highly charismatic, entertaining, and at times infuriating Socrates appeared to be a sorcerer bewitching the aristocratic young men of Athens who followed him fanatically about the agora.

By all credible accounts, this exceedingly eccentric, self-styled radical truth-seeker had more than a whiff of the uncanny about him. As Socrates himself explains in Plato's *Republic*, he was both blessed and burdened with a supernatural phenomenon in the form of a *daimonion* or inner spirit that always guided him: "This began when I was a child. It is a voice, and whenever it speaks, it turns me away from something I am about to do, but it never encourages me to do anything." An overtly rational thinker, Socrates nonetheless considered these warnings—or, in James Miller's words, "the audible interdictions he experienced as irresistible"—to be infallible. Such oracular injunctions were highly anomalous as tutelary spirits were thought to assume a more nuanced presence. Some scholars have dismissed Socrates' recurring sign as a hallucination or psychological aberration. Others have conjectured that the internal voice might be attributable to the cataleptic or trancelike episodes from which the philosopher purportedly suffered. Indeed, as Miller notes, "Socrates was storied for the abstracted states that overtook him"; not infrequently, his companions would see him stop in his tracks and stand still for hours, completely lost in thought.

As Socrates further insisted, it was only the protestations of this apotreptic voice that held him back from entering the political arena. Even so, its personal admonitions could not spare him persecution. Despite the political amnesty extended by the resurgent democracy that succeeded the interim pro-Spartan oligarchy, the thinker's notoriety and ambiguous allegiances aroused suspicions. In 399 BCE, Socrates was brought before the court on trumped-up charges of impiety; these included willfully neglecting the traditional divinities, flagrantly introducing new gods to the city, and wittingly corrupting the youth. Athenian society recognized no division between religious and civic duties, and capricious gods demanded constant appeasement through sacrifices and rituals. Consequently, belief in a purely private deity—particularly a wholly benevolent deity conveying unequivocal messages—was inadmissible. Worse, as Socrates' own testimony revealed, he honored this personal god's authority above even the laws of the city. Hence, the philosopher's *daimonion* loomed over his indictment, conviction, and sentencing.

Nevertheless, in his defense speech as reconstructed by Plato in the *Apology*, Socrates maintained confidence in the protective nature and prophetic powers of his inner monitor. He never questioned its affirmatory silence toward his predicament, remarking, "The divine faculty would surely have opposed me had I been going to evil and not to good." Thus, Socrates acknowledged that his *daimonion* had its reasons, however inscrutable. Variously described as malcontent and martyr, public nuisance and prophet, laughingstock and hero, the mercurial Athenian, like the sign that guided him, was difficult to fathom yet impossible to ignore.

*The Divine Sign of Socrates* ©UWorld

Annotations for this passage can be found in the CARS Passage Booklet.

### 2e. Determining Passage Perspectives Question 3

In describing Socrates as being "uncanny" and like "a sorcerer," the author seems to suggest that:

- A. only Socrates' followers could understand his method.
- B. Socrates tried to trick people into following him.
- C. Socrates was involved in magic or underhanded practices.
- D. Socrates' motivations were not easy for many people to discern.

See next page for the *strategy-based explanation* of this question.

# Chapter 6: Skill 2 – Reasoning Within the Text

> ## 2e. Determining Passage Perspectives Question 3
> ## Strategy-based Explanation
>
> In describing Socrates as being "uncanny" and like "a sorcerer," the author seems to suggest:
>
> A. that only Socrates' followers could understand his method.
> B. that Socrates tried to trick people into following him.
> C. that Socrates was involved in magic or underhanded practices.
> D. that Socrates' motivations were not easy for many people to discern.

Unlike the other 2e questions you have encountered so far, this question refers to specific viewpoint indicators and asks you to infer what the author is suggesting by their use. Accordingly, you will need to note these (and possibly other) viewpoint indicators in the context of the passage to determine a particular perspective of the author about Socrates.

## Applying the Method

*Passage Excerpt*

**Step 1: Take Note of Viewpoint Indicators**

[P1] Yet the philosopher's peculiar appearance was but a hint of the strange new calling he embraced. Inspired perhaps by the famous Delphic dictum "Know thyself," he embarked on a mission devoted to finding truth through dialogue. In what struck some as a dangerous new method of inquiry, he subjected nearly everyone he encountered to intense cross-examination, mercilessly exposing the ignorance of his interlocutors. Moreover, in a culture that still put stock in magic, the highly charismatic, entertaining, and at times infuriating Socrates **appeared to be a sorcerer** bewitching the aristocratic young men of Athens who followed him fanatically about the agora.

**Viewpoint Indicator**

In Paragraph 1, the author claims that Socrates **appeared to be a sorcerer** given how easily he drew in the aristocratic young men of Athens who followed him fanatically.

[P2] **By all credible accounts**, this exceedingly eccentric, self-styled radical truth-seeker **had more than a whiff of the uncanny about him.** As Socrates himself explains in Plato's *Republic*, he was both blessed and burdened with a supernatural phenomenon in the form of a *daimonion* or inner spirit that always guided him: "This began when I was a child. It is a voice, and whenever it speaks, it turns me away from something I am about to do, but it never encourages me to do

**Viewpoint Indicators**

By ascribing their view to **all credible accounts,** the author implies there is a consensus that Socrates **had more than a whiff of the uncanny about him**. In particular, Socrates claimed to be **guided by a supernatural spirit** that advised him. Although Socrates believed it was infallible, the spirit seemed **highly anomalous** to other people because it didn't fit with how spirits were typically thought of.

anything." An overtly rational thinker, Socrates nonetheless considered these warnings—or, in James Miller's words, "the audible interdictions he experienced as irresistible"—to be infallible. Such oracular injunctions were highly anomalous as tutelary spirits were thought to assume a more nuanced presence.

## Step 2: Determine the Source's Belief or Attitude

Having noted the viewpoint indicators in the context of the passage, we can now determine the author's perspective on Socrates. First, Socrates had such a strong sway over the young men who followed him that he **appeared to be a sorcerer** bewitching them. Thus, the citizens of Athens could not understand how Socrates inspired such devotion from his followers.

Similarly, Socrates **had more than a whiff of the uncanny about him**, partly because of his personal guiding spirit whose warnings he always obeyed. As the author further explains, the Athenians viewed this supposed spirit as "highly anomalous" because it didn't fit with the "more nuanced presence" such spirits were typically thought to have. Hence, this spirit constituted another aspect of Socrates that most Athenians did not understand.

Accordingly, both of these descriptions suggest that Socrates' behavior seemed confusing or abnormal to others. Therefore, the author's use of the terms "uncanny" and like "a sorcerer" seems to suggest **Choice D**, *that Socrates' motivations were not easy for many people to discern*.

Choice A, *that only Socrates' followers could understand his method*, may seem attractive as well, since it also concerns whether Socrates was understood. However, the "method of inquiry" referred to was a matter of questioning people and subjecting them to "intense cross-examination." While some viewed this method as "dangerous," the author gives no indication that people couldn't *understand* it. By contrast, the author does imply that people struggled to understand Socrates' motivations, making **Choice D** the best answer.

For an alternative method of explanation based more specifically on passage evidence, you can view this question in the UWorld Qbank (sold separately).

## Passage E: The Divine Sign of Socrates

From his bare feet to his bald pate, the potentially shapeshifting figure of Socrates found in the literary tradition that arose after his controversial trial and death presents an intriguing array of oddities and unorthodoxies. Most conspicuously, his unshod and shabby sartorial state flaunted poverty at a time when the city of Athens had become obsessed with wealth and its trappings. Yet the philosopher's peculiar appearance was but a hint of the strange new calling he embraced. Inspired perhaps by the famous Delphic dictum "Know thyself," he embarked on a mission devoted to finding truth through dialogue. In what struck some as a dangerous new method of inquiry, he subjected nearly everyone he encountered to intense cross-examination, mercilessly exposing the ignorance of his interlocutors. Moreover, in a culture that still put stock in magic, the highly charismatic, entertaining, and at times infuriating Socrates appeared to be a sorcerer bewitching the aristocratic young men of Athens who followed him fanatically about the agora.

By all credible accounts, this exceedingly eccentric, self-styled radical truth-seeker had more than a whiff of the uncanny about him. As Socrates himself explains in Plato's *Republic*, he was both blessed and burdened with a supernatural phenomenon in the form of a *daimonion* or inner spirit that always guided him: "This began when I was a child. It is a voice, and whenever it speaks, it turns me away from something I am about to do, but it never encourages me to do anything." An overtly rational thinker, Socrates nonetheless considered these warnings—or, in James Miller's words, "the audible interdictions he experienced as irresistible"—to be infallible. Such oracular injunctions were highly anomalous as tutelary spirits were thought to assume a more nuanced presence. Some scholars have dismissed Socrates' recurring sign as a hallucination or psychological aberration. Others have conjectured that the internal voice might be attributable to the cataleptic or trancelike episodes from which the philosopher purportedly suffered. Indeed, as Miller notes, "Socrates was storied for the abstracted states that overtook him"; not infrequently, his companions would see him stop in his tracks and stand still for hours, completely lost in thought.

As Socrates further insisted, it was only the protestations of this apotreptic voice that held him back from entering the political arena. Even so, its personal admonitions could not spare him persecution. Despite the political amnesty extended by the resurgent democracy that succeeded the interim pro-Spartan oligarchy, the thinker's notoriety and ambiguous allegiances aroused suspicions. In 399 BCE, Socrates was brought before the court on trumped-up charges of impiety; these included willfully neglecting the traditional divinities, flagrantly introducing new gods to the city, and wittingly corrupting the youth. Athenian society recognized no division between religious and civic duties, and capricious gods demanded constant appeasement through sacrifices and rituals. Consequently, belief in a purely private deity—particularly a wholly benevolent deity conveying unequivocal messages—was inadmissible. Worse, as Socrates' own testimony revealed, he honored this personal god's authority above even the laws of the city. Hence, the philosopher's *daimonion* loomed over his indictment, conviction, and sentencing.

Nevertheless, in his defense speech as reconstructed by Plato in the *Apology*, Socrates maintained confidence in the protective nature and prophetic powers of his inner monitor. He never questioned its affirmatory silence toward his predicament, remarking, "The divine faculty would surely have opposed me had I been going to evil and not to good." Thus, Socrates acknowledged that his *daimonion* had its reasons, however inscrutable. Variously described as malcontent and martyr, public nuisance and prophet, laughingstock and hero, the mercurial Athenian, like the sign that guided him, was difficult to fathom yet impossible to ignore.

*The Divine Sign of Socrates* ©UWorld

Annotations for this passage can be found in the CARS Passage Booklet.

## 2e. Determining Passage Perspectives Question 4

The passage author's view of Socrates' many eccentricities is that they constituted:

- A. a means of disguising deceptive practices.
- B. an essential part of his mission and message.
- C. a pattern of behavior characteristic of a political rebel.
- D. symptoms of an identifiable syndrome from which he suffered.

See next page for the *strategy-based explanation* of this question.

## 2e. Determining Passage Perspectives Question 4
## Strategy-based Explanation

The passage author's view of Socrates' many eccentricities is that they constituted:

A. a means of disguising deceptive practices.
B. an essential part of his mission and message.
C. a pattern of behavior characteristic of a political rebel.
D. symptoms of an identifiable syndrome from which he suffered.

This question asks you to determine the author's overall attitude about the various eccentricities that Socrates exhibited. While most of the passage is dedicated to just one of these eccentricities—Socrates' faith in his personal *daimonion*—you may recall that the passage begins by discussing other ways in which Socrates was unusual. By considering how all these aspects of Socrates are portrayed, we can determine the author's perspective about them.

## Applying the Method

*Passage Excerpt*

[P1] From his bare feet to his bald pate, the potentially shapeshifting figure of Socrates found in the literary tradition that arose after his controversial trial and death presents an intriguing array of oddities and unorthodoxies. Most conspicuously, his unshod and shabby sartorial state flaunted poverty at a time when the city of Athens had become obsessed with wealth and its trappings. Yet the philosopher's peculiar appearance was but a hint of the strange new calling he embraced. Inspired perhaps by the famous Delphic dictum "Know thyself," he embarked on a mission devoted to finding truth through dialogue. In what struck some as a dangerous new method of inquiry, he subjected nearly everyone he encountered to intense cross-examination, mercilessly exposing the ignorance of his interlocutors.

[P2] By all credible accounts, this exceedingly eccentric, self-styled radical truth-seeker had more than a whiff of the uncanny about him. As Socrates himself explains in Plato's *Republic*, he was both blessed and burdened with a supernatural phenomenon in the form of a *daimonion* or inner spirit that always guided him: "This began when I was a child. It

### Step 1: Take Note of Viewpoint Indicators

**Viewpoint Indicators**

Paragraph 1 describes what the author calls the **intriguing array of oddities and unorthodoxies** associated with Socrates. **Most conspicuously**, Socrates flaunted poverty while his society was obsessed with wealth. Thus, his behavior intentionally challenged society's values.

In addition, he set out on what **many saw as a strange** or even **dangerous** mission: mercilessly questioning everyone he met and exposing people's ignorance in a search for truth. Accordingly, this unusual practice also served to challenge people's beliefs.

**Viewpoint Indicators**

Paragraph 2 similarly emphasizes that **by all credible accounts**, Socrates was not only **exceedingly eccentric** but a **radical truth-seeker**. He further claimed that his unusual behavior was **guided by a supernatural spirit** that most people viewed as **highly anomalous**.

is a voice, and whenever it speaks, it turns me away from something I am about to do, but it never encourages me to do anything." An overtly rational thinker, Socrates nonetheless considered these warnings—or, in James Miller's words, "the audible interdictions he experienced as irresistible"—to be infallible. Such oracular injunctions were highly anomalous as tutelary spirits were thought to assume a more nuanced presence.

### Step 2: Determine the Source's Belief or Attitude

As we have seen, the author portrays Socrates' **intriguing array of oddities and unorthodoxies** as more than just unusual thoughts and behaviors. Rather, they are presented as part of Socrates' challenge of cultural norms. While most Athenians valued wealth, Socrates **conspicuously** flaunted poverty. While his method **struck people as dangerous**, he mercilessly exposed their ignorance. Thus, Socrates' **exceedingly eccentric** behavior was that of a **radical truth-seeker** who believed the Athenians should change their beliefs and values.

Accordingly, the author presents Socrates' eccentricities as closely intertwined with his dedication to finding truth and questioning his society. Therefore, in the author's view, Socrates' many eccentricities constituted **Choice B**, *an essential part of his mission and message*.

For an alternative method of explanation based more specifically on passage evidence, you can view this question in the UWorld Qbank (sold separately).

# Passage C: Probability and The Universe

The idea of probability is frequently misunderstood, in large part because of a conceptual confusion between objective probability and subjective probability. The failure to make this distinction leads to an erroneous conflation of genuine possibility with what is in fact merely personal ignorance of outcome. An example will clarify.

A standard die is rolled on a table, but the outcome of the roll is concealed. Should an observer be asked the chance that a particular number was rolled—five, say—the natural response is 1/6. However, this answer is incorrect. To say there is a one-in-six chance that a five was rolled implies there is an equal chance that any of the other numbers were rolled. But there is no equal chance, because the roll has already occurred. Hence, the probability that the result of the roll is a five is either 100% or 0%, and the same is true for each of the other numbers.

It might be objected that such an analysis is an issue of semantics rather than a substantive claim. For declaring the probability to be 1/6 is merely an expression that, for all we know, any number from 1 through 6 might have been rolled. But the difference between *for all we know* and what *is* remains crucially important, because it forestalls a tendency to make scientific assertions from a perspective biased by human perception....

[T]hus, it should be clear that claims about the purpose of the universe rest on shaky ground. In particular, we must be wary of inferences drawn from juxtaposing the existence of intelligent life with the genuinely improbable cosmological conditions that make such life possible. Joseph Zycinski provides an instructive account of those conditions: "If twenty billion years ago the rate of expansion were minimally smaller, the universe would have entered a stage of contraction at a time when temperatures of thousands of degrees were prevalent and life could not have appeared. If the distribution of matter were more uniform the galaxies could not have formed. If it were less uniform, instead of the presently observed stars, black holes would have formed from the collapsing matter." In short, the existence of life (let alone intelligent life) depends upon the initial conditions of the universe having conformed to an extremely narrow range of possible values.

It is tempting to go beyond Zycinski's factual point to draw deep cosmological, teleological, and even theological conclusions. But no such conclusions follow. The reason why a life-sustaining universe exists is that if it did not, there would be no one to wonder why a life-sustaining universe exists. This fact is a function not of purpose, but of pre-requisite. For instance, suppose again that a five was rolled on a die. We now observe a five on its face, not because the five was "meant to be" but because one side of the die had to land face up. Even if we imagine that the die has not six but six *billion* sides, that analysis is unaffected. Yes, it was unlikely that any given value would be rolled, but *some* value had to be rolled. That "initial condition" then set the parameters for what kind of events could possibly follow; in this case, the observation of the five.

Similarly, the existence of intelligent beings is evidence that certain physical laws obtained in the universe. However, it is not evidence that those beings or laws were necessary or intended rather than essentially random. The subjective probability of that outcome is irrelevant, because in this universe the objective probability of its occurrence is 100%. Thus, the seemingly low initial chance that such a universe would exist is not in itself indicative of a purpose to its existence.

*Probability and The Universe* ©UWorld

Annotations for this passage can be found in the CARS Passage Booklet.

### 2e. Determining Passage Perspectives Question 5

Based on the passage, with which of the following statements would the author most likely agree?

A. Some things must simply be accepted.
B. Math is the only universal language.
C. God does not play dice.
D. Everything happens for a reason.

See next page for the *strategy-based explanation* of this question.

Chapter 6: Skill 2 – Reasoning Within the Text

## 2e. Determining Passage Perspectives Question 5
### Strategy-based Explanation

Based on the passage, with which of the following statements would the author most likely agree?

A. Some things must simply be accepted.
B. Math is the only universal language.
C. God does not play dice.
D. Everything happens for a reason.

The question asks us to consider what the passage suggests about the author's perspective. Accordingly, we can expect to find one answer choice that clearly matches a view the author expresses, while the other choices are either contrary to that view or not discussed. In other words, the author's stated beliefs will confirm, refute, or be unrelated to each of the answer choices.

## Applying the Method

*Passage Excerpt*

**[P2]** To say there is a one-in-six chance that a five was rolled implies there is an equal chance that any of the other numbers were rolled. But there is no equal chance, because the roll has already occurred. **Hence**, the probability that the result of the roll is a five is either 100% or 0%, and the same is true for each of the other numbers.

**[P3]** But the difference between *for all we know* and what *is* remains **crucially important**, because it forestalls a tendency to make scientific assertions from a perspective biased by human perception....

**[P4]** [T]hus, **it should be clear that** claims about the purpose of the universe rest on shaky ground. In particular, we must be wary of inferences drawn from juxtaposing the existence of intelligent life with the genuinely improbable cosmological conditions that make such life possible.

**[P5]** It is tempting to go beyond Zycinski's factual point to draw deep cosmological, teleological, and even theological conclusions. But **no such conclusions follow**. The reason why a life-sustaining universe exists is that if it did not, there

### Step 1: Take Note of Viewpoint Indicators

**Viewpoint Indicator**

Choice B: Math is the only universal language.

Although the author draws conclusions about probability, they do not otherwise express a view about math, and they never compare math to a language.

**Viewpoint Indicator**

Choice A: Some things must simply be accepted.

In Paragraph 3 the author states that it is **crucially important** for scientific assertions to not be biased by human perception. In other words, the author thinks that facts about science must be accepted as they are, not distorted by human psychology.

Paragraph 6 applies that same view to a specific topic. The author states that facts about the universe are **not evidence that** the universe had a purpose; instead, it **may have just come about randomly**. Accordingly, the author thinks the existence of the universe might be a fact we have to simply accept without assuming any deeper explanation behind it.

would be no one to wonder why a life-sustaining universe exists. **This fact** is a function not of purpose, but of pre- requisite.

[P6] Similarly, the existence of intelligent beings is evidence that certain physical laws obtained in the universe. **However, it is not evidence that** those beings or laws were necessary or intended rather than essentially random. The subjective probability of that outcome is irrelevant, because in this universe the objective probability of its occurrence is 100%.

**Thus**, the seemingly low initial chance that such a universe would exist is not in itself indicative of a purpose to its existence.

---

**Viewpoint Indicator**

**Choice D: Everything happens for a reason.**

The author stresses multiple times that the universe **may not have a purpose**. Paragraph 4 states that this position **should be clear**; Paragraph 5 refers to that position as a **fact**; and Paragraph 6 **thus** draws the same conclusion about the universe.

**Viewpoint Indicator**

**Choice C: God does not play dice.**

The passage refers to God in relation to theological conclusions that one might be tempted to draw. However, the author then states that **no such conclusions follow**, revealing that they do not believe such conclusions about God are justified.

---

**Step 2: Determine the Source's Belief or Attitude**

Looking at the author's viewpoint indicators, we find a consistent message in the passage. According to the author, it is **crucially important** to avoid human bias in science; the improbability of the universe **is not evidence that** it had a purpose; and the universe may have just come about randomly. Thus, the author conveys that the existence of the universe may simply be a fact without any higher reason behind it. Accordingly, the answer choice that the author would most agree with is **Choice A**, *Some things must simply be accepted*.

Since the passage discusses how the universe came into existence, we might have initially been attracted to Choice D, *Everything happens for a reason*. However, reading attentively shows that this claim represents the *opposite* of the author's view; they argue that the universe and its life might have occurred randomly or for no particular reason. Thus, only **Choice A** accurately reflects the view expressed in the passage.

---

For an alternative method of explanation based more specifically on passage evidence, you can view this question in the UWorld Qbank (sold separately).

Lesson 6.6
# RWT Subskill 2f. Drawing Additional Inferences

| Skill 2: Reasoning Within the Text ||
|---|---|
| **Subskill** | **Student Objective** |
| 2a. Logical Relationships Within Passage | Identify logical relationships between passage claims |
| 2b. Function of Passage Claim | Determine what function a particular claim serves in the passage |
| 2c. Extent of Passage Evidence | Determine the extent to which the passage provides evidence for its claims |
| 2d. Connecting Claims With Evidence | Connect passage claims with supporting passage evidence |
| 2e. Determining Passage Perspectives | Determine the perspective of the author or another source in the passage |
| **2f. Drawing Additional Inferences** | **Draw additional inferences based on passage information** |

The final subskill for Reasoning Within the Text questions is **2f. Drawing Additional Inferences**. As mentioned in Lesson 5.6, these questions are similar to those in the **1f. Further Implications of Passage Claims** category. However, answering the RWT version of these questions may involve considering more of the passage or engaging in additional reasoning. Nevertheless, this process is still a matter of taking existing passage information and recognizing a further conclusion that can be drawn from it.

## Method for Answering 2f. Drawing Additional Inferences Questions

### Step 1: Verify What the Passage Directly States

As with **1f** questions, the first step in approaching **2f** questions is to review the direct passage claims that relate to the question topic. Typically, there will be multiple such claims that, when taken together, will suggest a new conclusion.

## Step 2: Synthesize the Relevant Information

The next step is to consider those claims as a group to determine what they collectively suggest. By synthesizing the relevant passage information in this way, you are drawing a new inference that either summarizes that information or logically follows from it. You can think of this process as forming a new understanding based on multiple points of data.

## Step 3: Compare Answer Choices with Your Inference

The last step is to compare the answer choices with the inference you have drawn. The correct answer will represent a claim that would fit with the existing passage information, either as a general restatement of it or as a closely related conclusion. Accordingly, this answer would sound natural if it were stated as an explicit claim in the passage.

## Subskill 2f. Drawing Additional Inferences
## (3 Practice Questions)

We now look at three examples from the UWorld Qbank. In each case, the passage and question are first given without commentary, allowing you to practice applying the method yourself. Then, the passage and question are presented again, this time with annotations of the passage (see the CARS Passage Booklet) and a step-by-step explanation of the question using the described method.

## Passage H: For Whom the Bell Toils

In nineteenth-century America, most people dismissed the notion that someone might assassinate the president. The presumption was based not only on ethics but practicality: a president's term is inherently limited, and an unpopular one could be voted out of office. Therefore, it was reasoned, there would be no need to consider removal through violence. This belief persisted even after the shocking murder of Abraham Lincoln in 1865, which was viewed as an aberration. Thus it was that on July 2, 1881, Charles Guiteau could simply walk up to President James A. Garfield and shoot him in broad daylight. As Richard Menke portrays events, "Guiteau was in fact a madman who had come to identify with a disgruntled wing of the Republican Party after his deranged fantasies of winning a post from the new administration had come to nothing." Believing that God had told him to kill the president, Guiteau thought this act would garner fame for his religious ideas and thereby help to usher in the Apocalypse.

In an interesting parallel, Garfield had felt a sense of divine purpose for his own life after surviving a near-drowning as a young man. Unlike Guiteau's fanatical ravings, however, Garfield's vision worked to the betterment of himself and the world. Candice Millard describes his ascent from extreme poverty to incredible excellence in college, where "by his second year…they made him a professor of literature, mathematics, and ancient languages." Garfield would go on to join the Union Army, where he attained the rank of major general and argued that black soldiers should receive the same pay as their white compatriots. While serving in the war he was nominated for the House of Representatives but accepted the seat only after President Lincoln declared that the country had more need of him as a congressman than as a general. The reluctant politician would later himself become president under similar circumstances, after multiple factions of a deadlocked Republican convention unexpectedly nominated him instead of their original candidates in 1880. An honest man who opposed corruption within the party, Garfield strove both to heal the fractures of the Civil War and to uphold the aims for which it was fought, until "the equal sunlight of liberty shall shine upon every man, black or white, in the Union."

Although Guiteau's bullet would ultimately dim this light for Garfield, the president actually survived the initial attack and for a time appeared headed for recovery. Tragically, however, the hubris shown by his main physician, Dr. Willard Bliss, would lead instead to weeks of prolonged suffering. None of the doctors who examined Garfield were able to locate the bullet, and its lingering presence—along with the unwashed hands of the doctors who probed for it—led to an infection. As the president's condition worsened, inventor Alexander Graham Bell attempted to adapt his patented telephone technology to locate foreign metal in the human body. Inspired by speculation that the bullet's electromagnetic properties might be detectable, Bell used his newly developed "Induction Balance" device to listen for the sounds of electrical interference he hoped would isolate the site of the bullet.

Unfortunately, Bell's searches were unsuccessful. Like Garfield's doctors, he had been looking for the bullet in the wrong area. Menke asserts that Bell's efforts "would probably have fallen short" regardless. However, other historians suggest that Dr. Bliss, unwilling to consider challenges to his original assessment, prevented Bell from more thoroughly searching the president's body. Certainly, Bliss ignored the advice and protestations of other physicians, even as Garfield continued to decline. With death imminent, Garfield asked to be taken to his seaside cottage, where he died on the 19th of September.

*For Whom the Bell Toils* ©UWorld

Annotations for this passage can be found in the CARS Passage Booklet.

## 2f. Drawing Additional Inferences Question 1

The author's description of Charles Guiteau (Paragraph 1) seems to imply that:

A. Guiteau viewed religion as a necessary tool for achieving his political ends.
B. Guiteau viewed achieving religious and political ends as essentially interconnected.
C. Guiteau pretended to hold religious views as an excuse for achieving his political ends through violence.
D. Guiteau pretended to hold religious and political views as an excuse for avenging a personal grievance.

See next page for the *strategy-based explanation* of this question.

Chapter 6: Skill 2 – Reasoning Within the Text

> ## 2f. Drawing Additional Inferences Question 1
> ### Strategy-based Explanations
>
> The author's description of Charles Guiteau (Paragraph 1) seems to imply that:
>
> A. Guiteau viewed religion as a necessary tool for achieving his political ends.
> B. Guiteau viewed achieving religious and political ends as essentially interconnected.
> C. Guiteau pretended to hold religious views as an excuse for achieving his political ends through violence.
> D. Guiteau pretended to hold religious and political views as an excuse for avenging a personal grievance.

The question specifies that the description of Charles Guiteau comes in Paragraph 1, so we can quickly find the necessary information to address it. The correct answer will be an inference about Guiteau that we can draw solely from the few claims the author makes about him.

## Applying the Method

*Passage Excerpt*

**[P1]** Thus it was that on July 2, 1881, Charles Guiteau could simply walk up to President James A. Garfield and shoot him in broad daylight. As Richard Menke portrays events, "Guiteau was in fact a madman who had come to identify with a disgruntled wing of the Republican Party after his deranged fantasies of winning a post from the new administration had come to nothing." Believing that God had told him to kill the president, Guiteau thought this act would garner fame for his religious ideas and thereby help to usher in the Apocalypse.

### Step 1: Verify What the Passage Directly States

The author's description of Charles Guiteau specifically relates to his assassination of President Garfield. According to this description, Guiteau was a madman in terms of both his political and religious views.

First, Guiteau had a political grievance toward Garfield that arose when his deranged fantasies of winning a post from the new administration failed to materialize.

Second, Guiteau had religious delusions: he believed God told him to kill the president and thought doing so would help to usher in the Apocalypse.

### Step 2: Synthesize the Relevant Information

In describing Charles Guiteau, the author focuses on two driving factors: Guiteau's political grievance and his religious delusions. Each of these factors appears to have motivated Guiteau's assassination of President Garfield.

Therefore, by considering these factors together, we can reasonably draw the following inference:

**Guiteau believed that killing Garfield would both avenge himself politically and further his religious goals.**

Accordingly, the correct answer choice should express something similar to this statement.

### Step 3: Compare Answer Choices with Your Inference

All the answer choices describe a relationship between Guiteau's political and religious views. However, the one that best matches our inference from Step 2 is **Choice B**, *Guiteau viewed achieving religious and political ends as essentially interconnected*. This statement aligns well with the author's description of Guiteau. Specifically, it reflects the fact that in Guiteau's mind, the same action that would give him revenge (**killing the president**) would also help bring about his religious goals (**the Apocalypse**).

Since Guiteau believed God had told him to kill Garfield, we might have been distracted by Choice A, *Guiteau viewed religion as a necessary tool for achieving his political ends*. However, the author never suggests that Guiteau used religion as a tool in the assassination. Rather, they say only that Guiteau "could simply walk up to" Garfield to shoot him. Moreover, this answer portrays Guiteau's religious views as subordinate to his political ones; by contrast, the passage depicts Guiteau as genuinely motivated by his religious delusions. Accordingly, Choice A is not a strong answer, whereas **Choice B** is well supported.

For an alternative method of explanation based more specifically on passage evidence, you can view this question in the UWorld Qbank (sold separately).

## Passage E: The Divine Sign of Socrates

From his bare feet to his bald pate, the potentially shapeshifting figure of Socrates found in the literary tradition that arose after his controversial trial and death presents an intriguing array of oddities and unorthodoxies. Most conspicuously, his unshod and shabby sartorial state flaunted poverty at a time when the city of Athens had become obsessed with wealth and its trappings. Yet the philosopher's peculiar appearance was but a hint of the strange new calling he embraced. Inspired perhaps by the famous Delphic dictum "Know thyself," he embarked on a mission devoted to finding truth through dialogue. In what struck some as a dangerous new method of inquiry, he subjected nearly everyone he encountered to intense cross-examination, mercilessly exposing the ignorance of his interlocutors. Moreover, in a culture that still put stock in magic, the highly charismatic, entertaining, and at times infuriating Socrates appeared to be a sorcerer bewitching the aristocratic young men of Athens who followed him fanatically about the agora.

By all credible accounts, this exceedingly eccentric, self-styled radical truth-seeker had more than a whiff of the uncanny about him. As Socrates himself explains in Plato's *Republic*, he was both blessed and burdened with a supernatural phenomenon in the form of a *daimonion* or inner spirit that always guided him: "This began when I was a child. It is a voice, and whenever it speaks, it turns me away from something I am about to do, but it never encourages me to do anything." An overtly rational thinker, Socrates nonetheless considered these warnings—or, in James Miller's words, "the audible interdictions he experienced as irresistible"—to be infallible. Such oracular injunctions were highly anomalous as tutelary spirits were thought to assume a more nuanced presence. Some scholars have dismissed Socrates' recurring sign as a hallucination or psychological aberration. Others have conjectured that the internal voice might be attributable to the cataleptic or trancelike episodes from which the philosopher purportedly suffered. Indeed, as Miller notes, "Socrates was storied for the abstracted states that overtook him"; not infrequently, his companions would see him stop in his tracks and stand still for hours, completely lost in thought.

As Socrates further insisted, it was only the protestations of this apotreptic voice that held him back from entering the political arena. Even so, its personal admonitions could not spare him persecution. Despite the political amnesty extended by the resurgent democracy that succeeded the interim pro-Spartan oligarchy, the thinker's notoriety and ambiguous allegiances aroused suspicions. In 399 BCE, Socrates was brought before the court on trumped-up charges of impiety; these included willfully neglecting the traditional divinities, flagrantly introducing new gods to the city, and wittingly corrupting the youth. Athenian society recognized no division between religious and civic duties, and capricious gods demanded constant appeasement through sacrifices and rituals. Consequently, belief in a purely private deity—particularly a wholly benevolent deity conveying unequivocal messages—was inadmissible. Worse, as Socrates' own testimony revealed, he honored this personal god's authority above even the laws of the city. Hence, the philosopher's *daimonion* loomed over his indictment, conviction, and sentencing.

Nevertheless, in his defense speech as reconstructed by Plato in the *Apology*, Socrates maintained confidence in the protective nature and prophetic powers of his inner monitor. He never questioned its affirmatory silence toward his predicament, remarking, "The divine faculty would surely have opposed me had I been going to evil and not to good." Thus, Socrates acknowledged that his *daimonion* had its reasons, however inscrutable. Variously described as malcontent and martyr, public nuisance and prophet, laughingstock and hero, the mercurial Athenian, like the sign that guided him, was difficult to fathom yet impossible to ignore.

*The Divine Sign of Socrates* ©UWorld

Annotations for this passage can be found in the CARS Passage Booklet.

### 2f. Drawing Additional Inferences Question 2

Information in the passage would suggest that Socrates' heavy influence on the aristocratic young men of ancient Athens was largely due to his:

  I. powerful personal magnetism.
  II. intense engagement in purposeful dialogue.
  III. intense devotion to civic duties.

A. I only
B. II only
C. I and II only
D. I, II, and III

See next page for the *strategy-based explanation* of this question.

## 2f. Drawing Additional Inferences Question 2
## Strategy-based Explanation

Information in the passage would suggest that Socrates' heavy influence on the aristocratic young men of ancient Athens was largely due to his:

   I. powerful personal magnetism.
   II. intense engagement in purposeful dialogue.
   III. intense devotion to civic duties.

   A. I only
   B. II only
   C. I and II only
   D. I, II, and III

The description of the aristocratic young men of Athens and their relationship with Socrates is limited to Paragraph 1. Accordingly, there is only a small amount of passage information needed to answer the question. These few claims will be sufficient to correctly infer why Socrates could exert such strong influence on the young men.

## Applying the Method

*Passage Excerpt*

**Step 1: Verify What the Passage Directly States**

[P1] From his bare feet to his bald pate, the potentially shapeshifting figure of Socrates found in the literary tradition that arose after his controversial trial and death presents an intriguing array of oddities and unorthodoxies. Most conspicuously, his unshod and shabby sartorial state flaunted poverty at a time when the city of Athens had become obsessed with wealth and its trappings. Yet the philosopher's peculiar appearance was but a hint of the strange new calling he embraced. Inspired perhaps by the famous Delphic dictum "Know thyself," he embarked on a **mission devoted to finding truth through dialogue**. In what struck some as a dangerous new method of inquiry, **he subjected nearly everyone he encountered to intense cross-examination**, mercilessly exposing the ignorance of his interlocutors. Moreover, in a culture that still put stock in magic, the **highly charismatic, entertaining, and at times infuriating Socrates** appeared to be a sorcerer **bewitching the aristocratic young men of Athens** who followed him fanatically about the agora.

Option II: intense engagement in purposeful dialogue.

In Paragraph 1, the author describes how Socrates engaged in a **mission devoted to finding truth** in which he constantly **subjected nearly everyone he encountered to intense cross-examination**. While Socrates engaged in this behavior, the aristocratic young men of Athens followed him fanatically about the agora, thus observing the process in action.

Option I: powerful personal magnetism.

Socrates himself is characterized as both **charismatic** and **entertaining**. Moreover, his influence over the young men of Athens was so great that he **appeared to be a sorcerer bewitching** them.

## Step 2: Synthesize the Relevant Information

The author calls Socrates **highly charismatic, entertaining**, and **bewitching** in describing his effect on the aristocratic young men of Athens. Therefore, we can reasonably infer that Socrates' influential personality attracted these young men to him. Accordingly, **Option I**, *powerful personal magnetism*, should be correct.

In addition, Socrates spent much of his time questioning everyone he met as part of a **mission devoted to finding truth through dialogue**. The author further tells us that the young men of Athens were so devoted to Socrates that they followed him fanatically about the agora. Since they clearly would have observed Socrates' behavior as they followed him, we can reasonably make the following inference: the young men of Athens enjoyed watching Socrates question people in search of truth. Therefore, **Option II**, *intense engagement in purposeful dialogue*, should also be correct.

However, the passage does not suggest anything about Option III, *intense devotion to civic duties*, in describing Socrates' effect on the young men of Athens. Thus, the correct answer should be *I and II only*.

## Step 3: Compare Answer Choices with Your Inference

After considering the relevant information in Paragraph 1, we inferred that Socrates' influence over the aristocratic young men of Athens was likely due to his highly charismatic personality and intense method of seeking the truth through dialogue, meaning the correct answer should be *I and II only*.

Looking at the answer choices, we see that this combination does appear, as **Choice C**. Therefore, we can feel confident that we have determined the correct answer.

For an alternative method of explanation based more specifically on passage evidence, you can view this question in the UWorld Qbank (sold separately).

## Passage I: Meaning: Readers or Authors?

Of late it has become popular among linguists and literary theorists to assert that a work's *meaning* depends upon the individual reader. It is readers, we are told, not authors, who create meaning, by interacting with a text rather than simply receiving it. Thus, a reader transcends the aims of the author, producing their own reading of the text. Indeed, on this line of thinking, even to speak of "the" text is to commit a conceptual error; every text is in fact many texts, a plurality of interpretations that resist comparative evaluation. This view is nonsense. That many otherwise sensible scholars should be attracted to it can perhaps be readily explained, but we should first delineate why the theory goes so far astray....

The absurdity of the view can be demonstrated by a practical analogy. Suppose Smith is conveying his ideas to Jones in conversation (the particular topic is of no consequence). Afterward, we discover that the men differ in their accounts of what Smith had expressed. At this point, Jones may decide that he misunderstood Smith, or perhaps that Smith was unclear. A more complex supposition might be that Smith misused some key term, so his words did not fully match his intentions. Any of these possibilities would reasonably describe why Jones and Smith possessed different opinions about what Smith had said.

What Jones may *not* justifiably conclude is that his own interpretation is what Smith *really meant*. He may not, in effect, say: "Yes, I admit that Smith honestly claims to have been saying something different, but I have formed my own equally correct understanding." Someone who made such an assertion would be suspected of making a joke; if he proved to be serious, we could only conclude that he was deeply confused or else being deliberately quarrelsome. For in questions about what Smith meant, it is surely Smith whose answer must be accepted…. [T]his is not a matter of *agreeing* with a speaker; Jones might judge Smith's ideas to be wrong, unfounded, etc. But whether Smith's ideas are right or wrong is a different matter from what those ideas *are*. On that count, Smith must be the authority.

However, this observation is in no way changed if Smith's ideas are written rather than spoken—sent by letter, for instance. Regardless of any interpretation Jones may produce, the letter's true meaning is whatever Smith intended to convey. Likewise it is, then, with a book, poem, or whatsoever object of literature a scholar (or ordinary reader) encounters. The writing down of ideas does not magically imbue them with malleability or render their content amorphous. From the loftiest tomes of Shakespeare or Milton to the lowliest of yellowed paperbacks, authors produce works with a particular message in mind. It is readers' task to discern that message, not to superimpose their own volitional perspectives.

To think otherwise is to undermine the foundation of literary scholarship. For what is the purpose of such scholarship, if not to seek understanding of an author's creation? One examines the text, taking note of style, historical context, allusions to other works, and other factors, in addition, of course, to the surface sense of the words themselves. If such an enterprise is to be reasonable, it must presume the existence of standards for success: accuracy and inaccuracy, depth or shallowness of analysis, grounds for preferring one interpretation to another. Different readers may come to different conclusions about a text, it is true. But to excise authorial intent from the evaluation of those conclusions does a disservice both to individual works and to literary study as a discipline.

*Meaning: Readers or Authors?* ©UWorld

Annotations for this passage can be found in the CARS Passage Booklet.

## 2f. Drawing Additional Inferences Question 3

The author's "more complex supposition" in Paragraph 2 implies that speakers:

- A. are not responsible for whether listeners correctly interpret the meaning of their ideas.
- B. cannot fail to accurately convey the meaning of their ideas.
- C. may sometimes be mistaken about the actual meaning of their ideas.
- D. can fail to accurately convey the meaning of their ideas.

See next page for the *strategy-based explanation* of this question.

> ### 2f. Drawing Additional Inferences Question 3
> ### Strategy-based Explanations
>
> The author's "more complex supposition" in Paragraph 2 implies that speakers:
>
> A. are not responsible for whether listeners correctly interpret the meaning of their ideas.
> B. cannot fail to accurately convey the meaning of their ideas.
> C. may sometimes be mistaken about the actual meaning of their ideas.
> D. can fail to accurately convey the meaning of their ideas.

The reference to a "more complex supposition" suggests a comparison with other suppositions or claims on the same topic. Accordingly, we can expect a set of related claims about speakers in Paragraph 2; these claims will provide the necessary information to determine what additional inference the supposition in question would justify.

## Applying the Method

*Passage Excerpt*

[P2] The absurdity of the view can be demonstrated by a practical analogy. Suppose Smith is conveying his ideas to Jones in conversation (the particular topic is of no consequence). Afterward, **we discover that the men differ in their accounts** of what Smith had expressed. At this point, Jones may decide that he misunderstood Smith, or perhaps that Smith was unclear. **A more complex supposition might be that Smith misused some key term, so his words did not fully match his intentions**. Any of these possibilities would **reasonably describe why** Jones and Smith possessed different opinions about what Smith had said.

### Step 1: Verify What the Passage Directly States

The author's **more complex supposition** is the last of three possible explanations for why Jones and Smith disagree about what Smith had said.

Possibility 1: Jones misunderstood Smith.

Possibility 2: Smith was unclear.

**Possibility 3: Smith misused some key term, so his words did not fully match his intentions.**

## Step 2: Synthesize the Relevant Information

If it were true that **Smith misused some key term, so his words did not fully match his intentions**, that would mean that he intended to say one thing but accidentally said something else. The author states that this possibility would **reasonably describe why** Jones and Smith possessed different opinions about what Smith had said. Based on this information, we can draw the following inference:

It is Smith's fault that he and Jones disagree about what Smith said.

However, the question doesn't ask what we can infer specifically about *Smith*; it asks what the scenario implies about speakers in general. Therefore, by extrapolating from Smith to any other speaker, our inference becomes:

**Sometimes it is the speaker's fault when listeners misunderstand them.**

Thus, the correct answer should state something similar to this claim.

## Step 3: Compare Answer Choices with Your Inference

Reviewing the answer choices, we see that **Choice D** is close to the inference we drew: speakers ***can fail to accurately convey the meaning of their ideas***. Just as Smith failed to use words that matched his intentions, any speaker might misuse a term or otherwise fail to convey what they really mean.

Choice C might seem to express something similar by stating that speakers *may sometimes be mistaken about the actual meaning of their ideas*. However, this answer choice confuses a mistake about *ideas* with a mistake about the meaning of *words*. In the author's scenario, Smith knows what he intends to say, or what his ideas are. His mistake involves misusing a key term, or thinking a word means something other than it actually does. Accordingly, Choice C does not reflect what the author's "more complex supposition" implies, and **Choice D** is the best answer.

For an alternative method of explanation based more specifically on passage evidence, you can view this question in the UWorld Qbank (sold separately).

Lesson 7.1

# RBT Subskill 3a. Exemplar Scenario for Passage Claims

The final competency or skill of the CARS section is **Reasoning Beyond the Text (RBT)**. These types of questions ask you to combine your understanding of the passage with information from outside it. Thus, although these questions incorporate external ideas, answering them still depends on passage information.

RBT questions can be classified into seven different types that correspond to specific subskills. For each subskill, a description is given, followed by a method for approaching such questions. Finally, example questions are provided to illustrate the method and allow you to practice.

| Skill 3: Reasoning Beyond the Text ||
| --- | --- |
| Subskill | Student Objective |
| 3a. Exemplar Scenario for Passage Claims | Identify which scenario most exemplifies or logically follows from passage claims |
| 3b. Passage Applications to New Context | Apply passage information to a new situation or context |
| 3c. New Claim Support or Challenge | Evaluate how a new claim supports or challenges passage information |
| 3d. External Scenario Support or Challenge | Determine which external scenario supports or challenges passage information |
| 3e. Applying Passage Perspectives | Apply the perspective of the author or another source in the passage to a new situation or claim |
| 3f. Additional Conclusions From New Information | Use new information to draw additional conclusions |
| 3g. Identifying Analogies | Identify analogies or similarities between passage ideas and ideas found outside the passage |

One of the most common types of Reasoning Beyond the Text questions is **3a. Exemplar Scenario for Passage Claims**. These questions ask you to determine which of the given options best corresponds with information found in the passage. For example, all of the following are **3a** questions:

- Which of the following scenarios is the best example of a "tyranny of chance" (Paragraph 6), as the passage author uses the term?

- If nineteenth-century society had embraced Fuller's ideas about gender, it would be reasonable to suppose that:

- Which of the following would be the best example of the passage author's claim that, in Voltaire's stories, humankind is portrayed as the cause of most of the suffering in the world?

- Based on Paragraphs 5 and 6, a modern lawyer would be most like Aristotle's "virtuous speaker" if she:

Thus, you can think of these questions as asking for the external example that best illustrates a particular passage claim or idea.

## Method for Answering 3a. Exemplar Scenario for Passage Claims Questions

### Step 1: Analyze the Relevant Passage Claims

Because answering these questions means finding the best example of passage claims, the first step is to analyze those claims to refresh your understanding of what you are looking for. This process may be more or less involved depending on the particular question. For instance, the wording of the earlier question about Voltaire's stories gives you a great deal of information, so reviewing the relevant passage material would likely be relatively quick. On the other hand, the question that refers to "Fuller's ideas about gender" provides no additional details, so it might require more analysis to verify what those ideas are.

### Step 2: Extrapolate to the Correct Answer

Once you have analyzed the relevant information, you can extrapolate from it to determine how the passage claims could best be exemplified. Although the answer choices all come from outside the passage, the correct answer should feel as though it could have been included as part of the passage itself. For example, suppose you encountered a question like:

"Which of the following fictional situations would best exemplify the passage claim that in medieval literature, every vow has a hidden danger?"

The correct answer might be: "*A man promises a witch his most valuable treasure in exchange for help, then is forced to give up not gold but his only child.*"

While this story and its characters might not actually appear in the passage, it is clear that such a situation would illustrate the claim being presented. Thus, while the correct answer does take you beyond the text, you can easily arrive at that answer through your analysis of passage claims.

> **Subskill Connection:**
> **1a. Main Idea or Purpose and 3a. Exemplar Scenario for Passage Claims**
>
> Step 2 may have reminded you of a point made in Lesson 5.1 about **1a. Main Idea or Purpose** questions. Specifically, that lesson stresses that even when a passage does not contain an explicit thesis statement, a correct description of the main idea would fit as such a statement if it had been included. In the same way, the correct answer to a **3a. Exemplar Scenario for Passage Claims** question should seem as if it could have been inserted into the relevant part of the passage following the words "for instance."

**Subskill 3a. Exemplar Scenario for Passage Claims
(3 Practice Questions)**

We now look at three examples from the UWorld Qbank. In each case, the passage and question are first given without commentary, allowing you to practice applying the method yourself. Then, the passage and question are presented again, this time with annotations of the passage (see the CARS Passage Booklet) and a step-by-step explanation of the question using the described method.

## Passage B: Jackie in 500 Words

Born in New York in 1929, Jacqueline Bouvier first came into the public eye as the wife of the 35th president of the United States, John F. Kennedy. The president was assassinated in 1963, and by the end of what turned out to be a turbulent decade, Mrs. Kennedy had transformed herself into the enigmatic Jackie O., wife of Greek shipping magnate Aristotle Onassis. Multifaceted and always elusive, the former first lady never ceased to fascinate; however, people had to be satisfied with only glimpses of this fashion icon, culture advocate, historic preservationist, polyglot, equestrienne, and book editor. Indeed, upon her death in 1994, Jacqueline Kennedy Onassis was described as "the most intriguing woman in the world." Often topping lists of the most admired individuals of the second half of the 20th century, this celebrated woman is likely someone many wish they had known. Barring such a possibility, the best way to fully appreciate Jackie's exceptional nature might be to consider the people she wished she had known.

It is unsurprising that this woman who captured the public's imagination for decades distinguished herself from her peers early on. Notably, in 1951, Ms. Bouvier entered a scholarship contest sponsored by *Vogue* and open to young women in their final undergraduate year, the annual Prix de Paris. Among other assignments, applicants were asked to compose a 500-word essay, "People I Wish I Had Known," spotlighting three individuals influential in art, literature, or culture. The future first lady chose an iconoclastic trio from the Victorian era: the French symbolist poet Charles Baudelaire, the Irish wit Oscar Wilde, and the innovative Ballets Russes dance company founder Sergei Diaghilev.

In a brief composition, Jackie provided deep insights into this bohemian threesome of poet, aesthete, and impresario with whom she strongly identified. She concluded that Baudelaire deployed "venom and despair" as "weapons" in his poetry. She idolized Wilde for being able "with the flash of an epigram to bring about what serious reformers had for years been trying to accomplish." Diaghilev she defined as an artist of a different sort, someone who "possessed what is rarer than artistic genius in any one field—the sensitivity to take the best of each man and incorporate it into a masterpiece." As Jackie poignantly observed, such a work is "all the more precious because it lives only in the minds of those who have seen it," dissipating soon after. Furthermore, although these men espoused different disciplines, she discerned that "a common theory runs through their work, a certain concept of the interrelation of the arts." Finally, foreshadowing her self-assumed role in the White House as the nation's unofficial minister of the arts, Jackie paid homage with her vision: "If I could be a sort of Overall Art Director of the Twentieth Century, watching everything from a chair hanging in space, it is their theories of art that I would apply to my period."

The contest committee judged Jackie's essay to have exhibited a profound appreciation for the arts combined with a truly outstanding level of intellectual maturity and originality of thought. Similarly, biographer Donald Spoto deemed Jackie "remarkably unorthodox," not unlike the men about whom she wrote in her unusual composition which he pronounced "a masterpiece of perceptive improvisation." Thus, from a pool of 1,279 applicants representing 224 colleges, Jacqueline Bouvier was declared the winner.

Although Ms. Bouvier went on to decline the prestigious award, which would have involved living and working in Paris, she never gave up her dream of being the century's art director. As first lady, she tirelessly promoted the arts and culture. Today, the John F. Kennedy Center for the Performing Arts in Washington, DC, is a legacy of Jackie's vision.

*Jackie in 500 Words* ©UWorld

Annotations for this passage can be found in the CARS Passage Booklet.

### 3a. Exemplar Scenario for Passage Claims Question 1

Which of the following significant accomplishments of Jackie's was most in accord with her stated aspiration?

    A. Promoting the establishment of a memorial library for the Kennedy presidency
    B. Organizing exhibitions of famous artworks and concerts by renowned musicians
    C. Setting fashion trends with elegant and sophisticated clothing styles
    D. Initiating a "historic restoration" of the White House and its furnishings

See next page for the *strategy-based explanation* of this question.

> ### 3a. Exemplar Scenario for Passage Claims Question 1
> ### Strategy-based Explanation
>
> Which of the following significant accomplishments of Jackie's was most in accord with her stated aspiration?
>
> A. Promoting the establishment of a memorial library for the Kennedy presidency
> B. Organizing exhibitions of famous artworks and concerts by renowned musicians
> C. Setting fashion trends with elegant and sophisticated clothing styles
> D. Initiating a "historic restoration" of the White House and its furnishings

This question is based on an aspiration of Jackie's that is discussed in the passage. Accordingly, determining the correct answer is a matter of identifying that aspiration and what it entails. You may recall that the passage begins by detailing Jackie's public profile before discussing her entry into the Prix de Paris scholarship essay contest. The description of that essay includes her aspiration or "vision" which we find quoted at the end of Paragraph 3.

## Applying the Method

*Passage Excerpt*

**[P3]** In a brief composition, Jackie provided deep insights into this bohemian threesome of poet, aesthete, and impresario with whom she strongly identified. She concluded that Baudelaire deployed "venom and despair" as "weapons" in his poetry. She idolized Wilde for being able "with the flash of an epigram to bring about what serious reformers had for years been trying to accomplish." Diaghilev she defined as an artist of a different sort, someone who "possessed what is rarer than artistic genius in any one field—the sensitivity to take the best of each man and incorporate it into a masterpiece." As Jackie poignantly observed, such a work is "all the more precious because it lives only in the minds of those who have seen it," dissipating soon after. Furthermore, although these men espoused different disciplines, she discerned that "a common theory runs through their work, a certain concept of the interrelation of the arts." Finally, foreshadowing her self-assumed role in the White House as the nation's unofficial minister of the arts, Jackie paid homage with her vision: "If I could be a sort of Overall Art Director of the Twentieth Century, watching everything from a chair

**Step 1: Analyze the Relevant Passage Claims**

Jackie's **aspiration** appears at the end of Paragraph 3, following a description of her essay and the theory of art she adopted.

From the men she admired and wrote about, she took on a concept of the arts as interrelated.

This view of the arts thus informed the **"vision"** or aspiration Jackie expressed: **"If I could be a sort of Overall Art Director of the Twentieth Century,** watching everything from a chair hanging in space, **it is their theories of art that I would apply to my period."**

hanging in space, **it is their theories of art that I would apply to my period."** | Accordingly, Jackie's stated aspiration was **to be an art director** who could **apply a theory of the arts as interrelated**.

### Step 2: Extrapolate to the Correct Answer

As we have seen, Jackie's stated aspiration was **to be an art director** who could **apply a theory of the arts as interrelated**. Therefore, the type of accomplishment that would exemplify this aspiration **would most reasonably involve promoting art events or portraying various kinds of art as related**.

Looking at the answer choices, we see that only **Choice B, *Organizing exhibitions of famous artworks and concerts by renowned musicians***, constitutes such an achievement: it shows Jackie acting as a director of events that made various types of art by famous artists available to the public. Therefore, **Choice B** reflects Jackie's stated aspiration and is the correct answer.

For an alternative method of explanation based more specifically on passage evidence, you can view this question in the UWorld Qbank (sold separately).

# Passage J: The Inkblots

For almost 100 years now, the psychological evaluation known as the Rorschach Inkblot Test has engendered much controversy, including skepticism about its value, questions about its scoring, and, especially, criticism of its interpretive methods as too subjective. Thus, the Rorschach test, which emerged from the same early twentieth-century zeitgeist that produced Einstein's physics, Freudian psychoanalysis, and abstract art, seems one of modernity's most misbegotten children. Destined never to be completely accepted or discredited, the test remains a perennial outlier in its field. Nevertheless, the inkblots' mystery and aesthetic appeal have caused them to be indelibly printed on our cultural fabric.

The now iconic inkblots were introduced to the world by Swiss psychiatrist Hermann Rorschach in his 1921 book *Psychodiagnostics*. As both director of the Herisau Asylum in Switzerland and an amateur artist, Rorschach was uniquely positioned to wed the new practice of psychoanalysis to the budding phenomenon of abstract art. For instance, reading Freud's work on dream symbolism prompted him to recall his childhood passion for a game based on inkblot art called *Klecksographie*. He was also cognizant that in a recently published dissertation, his colleague Szymon Hens had used inkblots to try to probe the imagination of research subjects; moreover, a few years earlier, the French psychologist and father of intelligence testing Alfred Binet had used them to measure creativity.

Motivated by these developments, the Herisau director decided to revisit that childhood pastime that had awakened his curiosity about how visual information is processed. In particular, he wondered why different people saw different things in the same image. Traditionally, psychoanalysts had relied on language for insights; however, as biographer Damion Searls reports, Rorschach's theories would exemplify the principle that "who we are is a matter less of what we say than of what we see." Indeed, through a process of perception termed pareidolia, the mind projects meaning onto images, detecting in them familiar objects or shapes. Consequently, what a person sees in an image reveals more about that person than about the image itself.

Rorschach experimented with countless inkblots, eventually selecting ten—five black on white, two also featuring some red, and three pastel-colored—to use with research subjects. For these perfectly symmetrical images—each of which he was said to have "meticulously designed to be as ambiguous and 'conflicted' as possible"—the primary question was always "What do you see?" Rorschach was especially careful to note how much attention individuals paid to various components of each inkblot (such as form, color, and a sense of movement) and whether they concentrated on details or the whole image. Having observed that his patients with schizophrenia gave distinctly different responses from the control group, Rorschach envisioned his experiment as a diagnostic tool for the disease.

Nevertheless, he resisted the notion that its results could be used to assess personality. In fact, until his untimely death from a ruptured appendix in 1922, Rorschach referred to his project as an "interpretive form experiment" rather than a test. Ironically, however, by the 1960s, the Rorschach Inkblot Test was known chiefly as a personality assessment and had become the most frequently administered projective personality test in the US.

Rorschach's test has survived nearly incessant scrutiny, including a 2013 comprehensive study of all Rorschach test data and repeated revisions to its scoring, yet doubts about its validity and reliability persist. Much like the inkblots themselves—which tantalize us with the possibility of divulging the secrets of who we are and how we see the world—the test has (for better or worse) defied attempts to fix its meaning. Thus, what has been called "the twentieth century's most visionary synthesis of art and science" stands tempered by harsh criticism.

*The Inkblots* ©UWorld

Annotations for this passage can be found in the CARS Passage Booklet.

## 3a. Exemplar Scenario for Passage Claims Question 2

Based on passage information, at the time of his sudden death in 1922, Rorschach was most likely intending:

- A. to continue his clinical research on schizophrenia.
- B. to adapt the inkblot test to diagnose other disorders.
- C. to publish a second volume of *Psychodiagnostics*.
- D. to develop an alternate method to assess personality.

See next page for the *strategy-based explanation* of this question.

Chapter 7: Skill 3 – Reasoning Beyond the Text

> ### 3a. Exemplar Scenario for Passage Claims Question 2
> ### Strategy-based Explanation
>
> Based on passage information, at the time of his sudden death in 1922, Rorschach was most likely intending:
>
> A. to continue his clinical research on schizophrenia.
> B. to adapt the inkblot test to diagnose other disorders.
> C. to publish a second volume of *Psychodiagnostics*.
> D. to develop an alternate method to assess personality.

This question could be rephrased as: "If he hadn't died unexpectedly, what would Rorschach most likely have done next?" One way to answer this question would be if the passage directly discusses Rorschach's intentions for the future. Alternatively, if no such intentions are discussed, we can assume that Rorschach would most likely have continued with whatever he was doing at the time of his death. You may recall that those activities are discussed in Paragraph 4.

## Applying the Method

*Passage Excerpt*

**Step 1: Analyze the Relevant Passage Claims**

[P4] Rorschach experimented with countless inkblots, eventually selecting ten—five black on white, two also featuring some red, and three pastel-colored—to use with research subjects. For these perfectly symmetrical images—each of which he was said to have "meticulously designed to be as ambiguous and 'conflicted' as possible"—the primary question was always "What do *you* see?" Rorschach was particularly careful to note how much attention individuals paid to various components of each inkblot (such as form, color, and a sense of movement) and whether they concentrated on details or the whole image. **Having observed that his patients with schizophrenia gave distinctly different responses from the control group, Rorschach envisioned his experiment as a diagnostic tool for the disease.** Nevertheless, he resisted the notion that its results could be used to assess personality. In fact, until his untimely death from a ruptured appendix in 1922, Rorschach referred to his project as an "interpretive form experiment" rather than a test. Ironically, however, by the 1960s, the Rorschach Inkblot Test was known chiefly as a personality assessment and

Paragraph 4 describes Rorschach's untimely death from a ruptured appendix in 1922 and discusses what he was doing at the time.

While conducting his inkblot experiment, Rorschach noticed that **patients with schizophrenia gave distinctly different responses** from the control group. Based on these results, he **envisioned his experiment becoming a diagnostic tool for the disease**.

Accordingly, the passage states that at the time of his sudden death, Rorschach was **studying patients with schizophrenia and considering**

255

had become the most frequently administered projective personality test in the US.

**how his inkblots might be used to help diagnose that condition**.

### Step 2: Extrapolate to the Correct Answer

We have determined that at the time of his sudden death, Rorschach was **studying patients with schizophrenia and considering how his inkblot experiment might be used to help diagnose the disease**. Therefore, it is reasonable to conclude that if he hadn't died, **Rorschach would have continued exploring whether his inkblots could be used to diagnose schizophrenia**.

Turning to the answer choices, we see that **Choice A**, *to continue his clinical research on schizophrenia*, reflects this conclusion. While the other answer choices may seem *possible*, the passage offers no reason to think any of them conveys Rorschach's actual intentions. By contrast, the passage does indicate that Rorschach was interested in exploring whether his inkblots could help diagnose schizophrenia. Accordingly, **Choice A** represents Rorschach's most likely intentions.

For an alternative method of explanation based more specifically on passage evidence, you can view this question in the UWorld Qbank (sold separately).

## Passage C: Probability and The Universe

The idea of probability is frequently misunderstood, in large part because of a conceptual confusion between objective probability and subjective probability. The failure to make this distinction leads to an erroneous conflation of genuine possibility with what is in fact merely personal ignorance of outcome. An example will clarify.

A standard die is rolled on a table, but the outcome of the roll is concealed. Should an observer be asked the chance that a particular number was rolled—five, say—the natural response is 1/6. However, this answer is incorrect. To say there is a one-in-six chance that a five was rolled implies there is an equal chance that any of the other numbers were rolled. But there is no equal chance, because the roll has already occurred. Hence, the probability that the result of the roll is a five is either 100% or 0%, and the same is true for each of the other numbers.

It might be objected that such an analysis is an issue of semantics rather than a substantive claim. For declaring the probability to be 1/6 is merely an expression that, for all we know, any number from 1 through 6 might have been rolled. But the difference between *for all we know* and what *is* remains crucially important, because it forestalls a tendency to make scientific assertions from a perspective biased by human perception….

[T]hus, it should be clear that claims about the purpose of the universe rest on shaky ground. In particular, we must be wary of inferences drawn from juxtaposing the existence of intelligent life with the genuinely improbable cosmological conditions that make such life possible. Joseph Zycinski provides an instructive account of those conditions: "If twenty billion years ago the rate of expansion were minimally smaller, the universe would have entered a stage of contraction at a time when temperatures of thousands of degrees were prevalent and life could not have appeared. If the distribution of matter were more uniform the galaxies could not have formed. If it were less uniform, instead of the presently observed stars, black holes would have formed from the collapsing matter." In short, the existence of life (let alone intelligent life) depends upon the initial conditions of the universe having conformed to an extremely narrow range of possible values.

It is tempting to go beyond Zycinski's factual point to draw deep cosmological, teleological, and even theological conclusions. But no such conclusions follow. The reason why a life-sustaining universe exists is that if it did not, there would be no one to wonder why a life-sustaining universe exists. This fact is a function not of purpose, but of pre-requisite. For instance, suppose again that a five was rolled on a die. We now observe a five on its face, not because the five was "meant to be" but because one side of the die had to land face up. Even if we imagine that the die has not six but six *billion* sides, that analysis is unaffected. Yes, it was unlikely that any given value would be rolled, but *some* value had to be rolled. That "initial condition" then set the parameters for what kind of events could possibly follow; in this case, the observation of the five.

Similarly, the existence of intelligent beings is evidence that certain physical laws obtained in the universe. However, it is not evidence that those beings or laws were necessary or intended rather than essentially random. The subjective probability of that outcome is irrelevant, because in this universe the objective probability of its occurrence is 100%. Thus, the seemingly low initial chance that such a universe would exist is not in itself indicative of a purpose to its existence.

*Probability and The Universe* ©UWorld

Annotations for this passage can be found in the CARS Passage Booklet.

> ### 3a. Exemplar Scenario for Passage Claims Question 3
>
> Which of the following scenarios best exemplifies "a conceptual confusion between objective probability and subjective probability" (Paragraph 1), as the author describes it?
>
> A. A researcher approximates the starting values for a computer simulation on population growth, runs the simulation twice, and is surprised to receive very different ending values each time.
>
> B. A law school graduate takes the bar exam, and while waiting for the results to be released she calculates how likely she is to pass.
>
> C. A gambler counts the number of aces that have been dealt, determines how many are left in the deck, and then judges whether another ace will be dealt.
>
> D. A hiker views a weather report predicting clear skies for the day, but when she is rained on during her hike, she realizes the meteorologist had been wrong.

See next page for the *strategy-based explanation* of this question.

## 3a. Exemplar Scenario for Passage Claims Question 3
## Strategy-based Explanation

Which of the following scenarios best exemplifies "a conceptual confusion between objective probability and subjective probability" (Paragraph 1), as the author describes it?

A. A researcher approximates the starting values for a computer simulation on population growth, runs the simulation twice, and is surprised to receive very different ending values each time.

B. A law school graduate takes the bar exam, and while waiting for the results to be released she calculates how likely she is to pass.

C. A gambler counts the number of aces that have been dealt, determines how many are left in the deck, and then judges whether another ace will be dealt.

D. A hiker views a weather report predicting clear skies for the day, but when she is rained on during her hike, she realizes the meteorologist had been wrong.

You may recall from reading the passage that it can be divided into two halves, the first of which focuses on explaining the conceptual confusion in question. By reviewing the author's distinction between objective and subjective probability in that half of the passage, we can see why someone would likely confuse those two concepts. Then, answering the question is just a matter of identifying the scenario that depicts a person being confused in that way.

## Applying the Method

*Passage Excerpt*

**[P1]** The idea of probability is frequently misunderstood, in large part because of a conceptual confusion between objective probability and subjective probability. The failure to make this distinction leads to an erroneous conflation of genuine possibility with what is in fact merely personal ignorance of outcome. An example will clarify.

**[P2]** A standard die is rolled on a table, but the outcome of the roll is concealed. Should an observer be asked the chance that a particular number was rolled—five, say—the natural response is 1/6. However, this answer is incorrect. To say there is a one-in-six chance that a five was rolled implies there is an equal chance that any of the other numbers were rolled. But there is no equal chance, because the roll has already occurred. Hence, the probability that the result of the roll is

### Step 1: Analyze the Relevant Passage Claims

In Paragraph 1, the author makes the following distinction between objective and subjective probability:

- objective probability is based on **genuine possibility**.
- subjective probability is based on **personal ignorance of outcome**.

Paragraph 2 then illustrates that distinction with an example:

- If you **don't know** what number was rolled on a die, then **you think** (subjectively) that the chance of a five is **1/6**.
- But since the roll has already happened, **in fact** (objectively) there is no "chance," the probability that a five was rolled is **either 100% or 0%**.

**a five is either 100% or 0%**, and the same is true for each of the other numbers.

**[P3]** It might be objected that such an analysis is an issue of semantics rather than a substantive claim. For declaring the probability to be 1/6 is merely an expression that, for all we know, any number from 1 through 6 might have been rolled. But the difference between *for all we know* and *what is* remains crucially important, because it forestalls a tendency to make scientific assertions from a perspective biased by human perception

Finally, Paragraph 3 sums up the distinction as:

- the difference between *for all we know* and *what is*.

Therefore, the conceptual confusion the author describes comes when:

- **people's subjective ignorance of outcome prevents them from realizing a past event is objectively no longer a matter of chance.**

### Step 2: Extrapolate to the Correct Answer

We have determined that the conceptual confusion between objective probability and subjective probability arises when **people's subjective ignorance of outcome prevents them from realizing a past event is objectively no longer a matter of chance**. Thus, we are looking for the answer choice that represents this same type of confusion.

In addition, since the author presents the die-rolling example as an illustration of this confusion, we could also approach the answer choices by asking: which scenario might the author have used instead of the die roll to make the same point?

Reviewing those scenarios, we can see that only one describes a person's subjective ignorance about a past event: **Choice B**, *A law school graduate takes the bar exam, and while waiting for the results to be released she calculates how likely she is to pass*. Since the graduate has already taken the exam, objectively she has already passed or failed—even though, from her subjective point of view, she does not yet know which it is. Thus, her case is like the die roll: the results of her exam are "concealed" from her, but they have still already been determined.

Therefore, **Choice B** best exemplifies the "conceptual confusion between objective probability and subjective probability" the author describes.

For an alternative method of explanation based more specifically on passage evidence, you can view this question in the UWorld Qbank (sold separately).

Lesson 7.2

# RBT Subskill 3b. Passage Applications to New Context

| Skill 3: Reasoning Beyond the Text ||
|---|---|
| **Subskill** | **Student Objective** |
| 3a. Exemplar Scenario for Passage Claims | Identify which scenario most exemplifies or logically follows from passage claims |
| **3b. Passage Applications to New Context** | **Apply passage information to a new situation or context** |
| 3c. New Claim Support or Challenge | Evaluate how a new claim supports or challenges passage information |
| 3d. External Scenario Support or Challenge | Determine which external scenario supports or challenges passage information |
| 3e. Applying Passage Perspectives | Apply the perspective of the author or another source in the passage to a new situation or claim |
| 3f. Additional Conclusions From New Information | Use new information to draw additional conclusions |
| 3g. Identifying Analogies | Identify analogies or similarities between passage ideas and ideas found outside the passage |

**3b. Passage Applications to New Context** questions introduce a new scenario to which you must apply information from the passage. Accordingly, such questions may ask:

- **How the new scenario illustrates claims from the passage**

    "In one episode of a 1981 TV show, an art critic praises a painting of a single circle, stating that it 'reduces experience to its most elemental aesthetic' and is 'quite profound in its simplicity.' This scenario best supports the assertion that:"

- **What passage information would suggest about the new scenario**

    "Suppose that the plays 'Shogun' and 'The King and I' are being performed at a Broadway theater in New York City. Based on the passage, which of the following is *least* likely to be true?"

- **How a situation described in the passage could hypothetically have been different**

    "Based on the passage, the need to tear down and rebuild the White House in the mid-twentieth century might have been prevented if:"

In many ways, these questions can be seen as the mirror image of **3a. Exemplar Scenario for Passage Claims** questions (see Lesson 7.1). Whereas **3a** questions ask which of the given scenarios best exemplifies passage information, **3b** questions provide a specific scenario and ask how passage information relates to it. Similarly, the method for approaching these questions can also be seen as a mirror image.

## Method for Answering 3b. Passage Applications to New Context Questions

### Step 1: Determine the Point of Connection

The key to answering **3b** questions is to identify the common element between the passage and the new information introduced by the question. There will always be a point of overlap between the two, or else the question would not make sense to ask. Accordingly, by analyzing the question scenario you can identify where that overlap could occur.

For example, the first question listed earlier describes *a TV art critic who praises a noticeably simple painting*. This scenario would likely connect to passage claims about one or more of the following:

- the views of art experts or connoisseurs;
- public perceptions or depictions of such experts;
- how paintings (or artworks in general) are judged;
- the simplicity or complexity of an artwork.

This does not mean that all these topics necessarily appear in the actual passage; rather, they constitute the kinds of topics that could logically connect to the TV art critic described. In addition, since you would have read the passage before considering that question, you would know which of those topics were actually discussed. Thus, after analyzing the question scenario, you could compare it with the relevant passage claims to determine their point of connection.

### Step 2: Apply That Connection to the Answer Choices

Having determined that point of connection, you know how the information introduced in the question scenario relates to claims made in the passage. Only one answer choice will accurately reflect the common element between them, revealing that answer to be correct.

## Subskill 3b. Passage Applications to New Context
### (3 Practice Questions)

We now look at three examples from the UWorld Qbank. In each case, the passage and question are first given without commentary, allowing you to practice applying the method yourself. Then, the passage and question are presented again, this time with annotations of the passage (see the CARS Passage Booklet) and a step-by-step explanation of the question using the described method.

## Passage G: Food Costs and Disease

Because frequent consumption of unhealthy foods is strongly linked with cardiometabolic diseases, one way for governments to combat those afflictions may be to modify the eating habits of the general public. Applying economic incentives or disincentives to various types of foods could potentially alter people's diets, leading to more positive health outcomes.

Utilizing national data from 2012 regarding food consumption, health, and economic status, Peñalvo et al. concluded that such price adjustments would help to prevent deaths related to cardiometabolic diseases. According to their analysis, increasing the prices of unhealthy foods such as processed meats and sugary sodas by 10%, while reducing the prices of healthy foods such as fruit and vegetables by 10%, would prevent an estimated 3.4% of yearly deaths in the U.S. Changing prices by 30% would have an even stronger effect, preventing an estimated 9.2% of yearly deaths. This data comports with that found in other countries, such as "previous modeling studies in South Africa and India, where a 20% SSB [sugar-sweetened beverage] tax was estimated to reduce diabetes prevalence by 4% over 20 years." The effects of price adjustments would be most pronounced on persons of lower socioeconomic status, as the researchers "found an overall 18.2% higher price-responsiveness for low versus high SES [socioeconomic status] groups."

This differential effect based on socioeconomic status contributes to concerns about such interventions, however. In *Harvard Public Health Review*, Kates and Hayward ask: "Well intentioned though they may be, at what point do these taxes overstep government influence on an individual's right to autonomy in decision-making? On whom does the increased financial burden of this taxation fall?" They note that taxes on sugar-sweetened beverages, for instance, "are likely to have a greater impact on low-income individuals…because individuals in those settings are more likely to be beholden to cost when making decisions about food."

However, "well intentioned though they may be," the worries that Kates and Hayward express are to some extent misguided. In particular, the idea that taxing unhealthy foods would burden those least able to afford it misses the point. Although the increased taxes would affect anyone who continued to purchase the items despite the higher prices, the goal of raising prices on unhealthy foods is precisely to dissuade people from buying them. As Kates and Hayward themselves remark, "Those in low-income environments may also be the largest consumers of obesogenic foods and therefore most likely to benefit from such a lifestyle change indirectly posed by SSB taxes." As the goal of the taxes is to promote those lifestyle changes, the financial burden objection is a non-starter.

Given this recognition, the question regarding autonomy constitutes a more substantial issue. Nevertheless, that concern also rests on a dubious assumption, as people's autonomy is not necessarily respected in the current situation either. The fact that those of lower socioeconomic status are more likely to have poorer diets suggests that such persons' food choices are the result of financial constraint, not fully autonomous, rational deliberation. Hence, by making healthy foods more affordable relative to unhealthy ones, government intervention might actually *facilitate* autonomous choices rather than hindering them.

On the other hand, suppose that the disproportionate consumption of unhealthy foods—and associated higher incidence of disease—among certain groups is not the result of financial hardship but rather the result of those persons' perceived self-interest. If so, that would suggest that members of these groups are being encouraged to persist in harmful dietary habits for the sake of corporate profits. In that case, violating autonomy for the sake of health may be permissible, as that practice would be morally preferable to the present system of corporate exploitation.

*Food Costs and Disease* ©UWorld

Annotations for this passage can be found in the CARS Passage Booklet.

### 3b. Passage Applications to New Context Question 1

A government worries that if price adjustments on food are implemented, citizens' autonomy will be violated to an unacceptable degree. Which recommendation represents the most reasonable compromise between the government's worry and the author's public health goals?

A. Explain planned price adjustments to the public in advance of their implementation.
B. Use nutritional education to shift consumers toward healthier foods without price adjustments.
C. Exclude sugar-sweetened beverages from price increases to reduce disproportionate impact on certain groups.
D. Reduce prices on healthy foods without increasing prices on unhealthy foods.

See next page for the *strategy-based explanation* of this question.

Chapter 7: Skill 3 – Reasoning Beyond the Text

### 3b. Passage Applications to New Context Question 1
### Strategy-based Explanation

A government worries that if price adjustments on food are implemented, citizens' autonomy will be violated to an unacceptable degree. Which recommendation represents the most reasonable compromise between the government's worry and the author's public health goals?

A. Explain planned price adjustments to the public in advance of their implementation.
B. Use nutritional education to shift consumers toward healthier foods without price adjustments.
C. Exclude sugar-sweetened beverages from price increases to reduce disproportionate impact on certain groups.
D. Reduce prices on healthy foods without increasing prices on unhealthy foods.

In the question scenario, a government believes that achieving the author's public health goals risks violating the autonomy of the citizens. However, the possibility of a "reasonable compromise" implies that this threat to autonomy could be reduced or removed while some progress would still be made toward the author's goals. Therefore, to identify this compromise we need to ask: what are the author's goals in the passage, and how do they connect to the government's worry in the question?

## Applying the Method

*Passage Excerpt*

**Step 1: Determine the Point of Connection**

[P1] Because frequent consumption of unhealthy foods is strongly linked with cardiometabolic diseases, one way for governments to combat those afflictions may be to modify the eating habits of the general public. Applying economic incentives or disincentives to various types of foods could potentially **alter people's diets, leading to more positive health outcomes.**

Paragraphs 1 and 2 explain the author's public health goals:

- Make healthy foods less expensive and unhealthy foods more expensive; which would
- **alter people's diets**, which would
- **reduce deaths from cardiometabolic disease**.

[P2] Utilizing national data from 2012 regarding food consumption, health, and economic status, Peñalvo et al. concluded that such price adjustments would help to prevent deaths related to cardiometabolic diseases. According to their analysis, increasing the prices of unhealthy foods such as processed meats and sugary sodas by 10%, while reducing the prices of healthy foods such as fruit and vegetables by 10%, **would prevent an estimated 3.4% of yearly deaths in the U.S**. Changing prices by 30% would have an even stronger effect, preventing an estimated 9.2% of yearly deaths.

Chapter 7: Skill 3 – Reasoning Beyond the Text

**[P3]** This differential effect based on socioeconomic status contributes to concerns about such interventions, however. In *Harvard Public Health Review,* Kates and Hayward ask: "Well intentioned though they may be, **at what point do these taxes overstep government influence on an individual's right to autonomy in decision-making?** On whom does the increased financial burden of this taxation fall?" They note that taxes on sugar-sweetened beverages, for instance, "are likely to have a greater impact on low-income individuals…because individuals in those settings are more likely to be beholden to cost when making decisions about food."

**[P4]** In particular, the idea that taxing unhealthy foods would burden those least able to afford it misses the point. Although the increased taxes would affect anyone who continued to purchase the items despite the higher prices, **the goal of raising prices on unhealthy foods is precisely to dissuade people from buying them**.

Paragraphs 3 and 4 then explain the government's worry in the question scenario:

- The author's proposal would **interfere with an individual's right to autonomy in decision-making**, because
- the increased taxes on unhealthy foods are designed **to inhibit people from buying those foods** even if they want to.

Therefore, the point of connection between the concern about autonomy in the question and the author's goals in the passage is the **increased taxes on unhealthy foods** which lead to the government's worry.

### Step 2: Apply That Connection to the Answer Choices

We determined that the connection between the author's public health goals and violating citizens' autonomy depends on the proposal to **increase taxes on unhealthy foods**. Thus, doing away with that proposal should also remove the threat to autonomy. As we look at the answer choices, we see that **Choice D** makes just that suggestion: *Reduce prices on healthy foods without increasing prices on unhealthy foods*.

This course of action would eliminate the coercive aspect of the author's plan, as the government would not be attempting to change consumers' behavior by penalizing the choice to buy unhealthy foods. However, it would preserve the other aspect of the author's plan, the suggestion to lower the price of healthy foods. This part of the plan could still influence consumers' behavior, but it would do so by *incentivizing the choice* to buy healthy foods, rather than trying to *inhibit the choice* to buy unhealthy foods.

Accordingly, this recommendation could help to achieve the author's public health goals, but without violating consumers' autonomy. Therefore, it represents the kind of reasonable compromise we are looking for, making **Choice D** the correct answer.

For an alternative method of explanation based more specifically on passage evidence, you can view this question in the UWorld Qbank (sold separately).

# Passage E: The Divine Sign of Socrates

From his bare feet to his bald pate, the potentially shapeshifting figure of Socrates found in the literary tradition that arose after his controversial trial and death presents an intriguing array of oddities and unorthodoxies. Most conspicuously, his unshod and shabby sartorial state flaunted poverty at a time when the city of Athens had become obsessed with wealth and its trappings. Yet the philosopher's peculiar appearance was but a hint of the strange new calling he embraced. Inspired perhaps by the famous Delphic dictum "Know thyself," he embarked on a mission devoted to finding truth through dialogue. In what struck some as a dangerous new method of inquiry, he subjected nearly everyone he encountered to intense cross-examination, mercilessly exposing the ignorance of his interlocutors. Moreover, in a culture that still put stock in magic, the highly charismatic, entertaining, and at times infuriating Socrates appeared to be a sorcerer bewitching the aristocratic young men of Athens who followed him fanatically about the agora.

By all credible accounts, this exceedingly eccentric, self-styled radical truth-seeker had more than a whiff of the uncanny about him. As Socrates himself explains in Plato's *Republic*, he was both blessed and burdened with a supernatural phenomenon in the form of a *daimonion* or inner spirit that always guided him: "This began when I was a child. It is a voice, and whenever it speaks, it turns me away from something I am about to do, but it never encourages me to do anything." An overtly rational thinker, Socrates nonetheless considered these warnings—or, in James Miller's words, "the audible interdictions he experienced as irresistible"—to be infallible. Such oracular injunctions were highly anomalous as tutelary spirits were thought to assume a more nuanced presence. Some scholars have dismissed Socrates' recurring sign as a hallucination or psychological aberration. Others have conjectured that the internal voice might be attributable to the cataleptic or trancelike episodes from which the philosopher purportedly suffered. Indeed, as Miller notes, "Socrates was storied for the abstracted states that overtook him"; not infrequently, his companions would see him stop in his tracks and stand still for hours, completely lost in thought.

As Socrates further insisted, it was only the protestations of this apotreptic voice that held him back from entering the political arena. Even so, its personal admonitions could not spare him persecution. Despite the political amnesty extended by the resurgent democracy that succeeded the interim pro-Spartan oligarchy, the thinker's notoriety and ambiguous allegiances aroused suspicions. In 399 BCE, Socrates was brought before the court on trumped-up charges of impiety; these included willfully neglecting the traditional divinities, flagrantly introducing new gods to the city, and wittingly corrupting the youth. Athenian society recognized no division between religious and civic duties, and capricious gods demanded constant appeasement through sacrifices and rituals. Consequently, belief in a purely private deity—particularly a wholly benevolent deity conveying unequivocal messages—was inadmissible. Worse, as Socrates' own testimony revealed, he honored this personal god's authority above even the laws of the city. Hence, the philosopher's *daimonion* loomed over his indictment, conviction, and sentencing.

Nevertheless, in his defense speech as reconstructed by Plato in the *Apology*, Socrates maintained confidence in the protective nature and prophetic powers of his inner monitor. He never questioned its affirmatory silence toward his predicament, remarking, "The divine faculty would surely have opposed me had I been going to evil and not to good." Thus, Socrates acknowledged that his *daimonion* had its reasons, however inscrutable. Variously described as malcontent and martyr, public nuisance and prophet, laughingstock and hero, the mercurial Athenian, like the sign that guided him, was difficult to fathom yet impossible to ignore.

*The Divine Sign of Socrates* ©UWorld

Annotations for this passage can be found in the CARS Passage Booklet.

### 3b. Passage Applications to New Context Question 2

In the *Theages*, an ancient dialogue traditionally attributed to Plato, a student asks Socrates whether he ought to train for an athletic competition. Given the information in the passage, which of the following would best explain why the student would have sought Socrates' counsel?

- A. Socrates was a highly eccentric character.
- B. Socrates' behavior had made him a well-known figure in Athenian society.
- C. Socrates was dedicated to discovering the truth through deliberative conversation.
- D. Socrates wore shabby clothing and was unconcerned with the appearance of wealth.

See next page for the *strategy-based explanation* of this question.

Chapter 7: Skill 3 – Reasoning Beyond the Text

## 3b. Passage Applications to New Context Question 2
## Strategy-based Explanation

In the *Theages*, an ancient dialogue traditionally attributed to Plato, a student asks Socrates whether he ought to train for an athletic competition. Given the information in the passage, which of the following would best explain why the student would have sought Socrates' counsel?

- A. Socrates was a highly eccentric character.
- B. Socrates' behavior had made him a well-known figure in Athenian society.
- C. Socrates was dedicated to discovering the truth through deliberative conversation.
- D. Socrates wore shabby clothing and was unconcerned with the appearance of wealth.

In this question scenario, a student seeks advice from Socrates, and we are asked to determine why he would do so. We could thus approach this question by asking: "Based on the passage, why would a student think Socrates would be a good person to ask for advice?"

## Applying the Method

*Passage Excerpt*

**[P1]** Inspired perhaps by the famous Delphic dictum "Know thyself," **he embarked on a mission devoted to finding truth through dialogue**. In what struck some as a dangerous **new method of inquiry**, he subjected nearly everyone he encountered to **intense cross-examination**, mercilessly exposing the ignorance of his interlocutors. Moreover, in a culture that still put stock in magic, the highly charismatic, entertaining, and at times infuriating Socrates appeared to be a sorcerer bewitching the aristocratic young men of Athens who followed him fanatically about the agora.

### Step 1: Determine the Point of Connection

Paragraph 1 describes Socrates' personal mission in life as well as the admiration he inspired in the young men of Athens.

Using **a new method of inquiry** involving **intense cross-examination**, Socrates would question nearly everyone he met. This mission, as witnessed and admired by the young men who followed him around fanatically, was dedicated to **finding truth through dialogue.**

Accordingly, the question scenario of the student seeking advice and the passage claims about Socrates are connected in terms of Socrates' dedication to finding truth and the young men's admiration for him.

## Step 2: Apply That Connection to the Answer Choices

As we determined in Step 1, the question scenario in which the student seeks advice and the passage claims about Socrates are connected in terms of both Socrates' dedication to finding truth and the young men's admiration for him. Turning to the answer choices, we see that **Choice C**, *Socrates was dedicated to discovering the truth through deliberative conversation*, best reflects these points of connection.

Since Socrates not only had a reputation for finding the truth but had also inspired the admiration of the young men who witnessed his method, it makes sense that a student would seek his guidance in making a decision. Accordingly, **Choice C** is the correct answer.

For an alternative method of explanation based more specifically on passage evidence, you can view this question in the UWorld Qbank (sold separately).

# Passage L: When Defense Is Indefensible

Suppose a prosecutor is considering whether to bring a case to trial. He is not sure that the suspect is guilty—in fact, based on the evidence, it's more likely that the suspect is *not* guilty. Nevertheless, he feels confident he can secure a guilty verdict. His powers of persuasion are considerable, and there's a good chance he could trick a jury into believing the evidence is strong instead of weak. In addition, the case is high profile and could be very lucrative; winning would likely lead to a substantial raise or promotion. He decides to charge the suspect, and ultimately succeeds in persuading the jury to convict.

Looking at this situation, most of us would easily judge the prosecutor as extremely unethical. His conduct is outrageous and wrong—he clearly acted with corrupt intent, perpetrating injustice in order to profit financially. Why is it not shocking, then, that we tolerate the mirror image of this behavior from defense attorneys? For they engage in the same outrageous conduct, only on the other side. Paid handsomely to represent even the vilest of clients, they apply their oratorical prowess to manipulating jury perception, keeping the guilty free and unpunished in exchange for money and status. To the extent that this behavior takes place, are some defense attorneys as unethical as our hypothetical prosecutor?

It is worth distinguishing two senses of the word "ethical" here. For there are standards of *professional* ethics to which any attorney must conform, including standards particular to the defense. Most relevant to our purposes, a lawyer is obligated to provide their client with a "zealous defense." In other words, once an attorney takes on a client, they are ethically bound to promote that client's rights, interests, or innocence—in fact, *not* to do so would be *unethical*. Thus, one might try to suggest that this obligation undermines the claim that some defense attorneys act unethically.

However, meeting that professional standard is not the same as being ethical in the general sense of the word. The standard depends on the condition: *once an attorney takes on a client*. With the exception of court-appointed attorneys or public defenders, who are assigned to provide representation to those who would otherwise lack it, an attorney is never required to represent a defendant. Therefore, meeting one's obligations as a defense attorney does not necessarily make one ethical, because the choice to accept a specific case (and thus to incur those obligations in the first place) may itself be an unethical act.

Moreover, the role of court-appointed attorneys is to help protect the rights of citizens who cannot secure their own representation, usually for financial reasons. Although preserving those rights is necessary to uphold justice, this situation highlights how wealth and class contribute to *injustice*. While some defendants possess the means to hire top-level private lawyers, others must depend on public servants—frequently less experienced lawyers from overloaded, understaffed agencies. As a result, the rich are more likely to escape conviction even when they are guilty, and the poor are more likely to be convicted even when they are innocent.

It is doubtful that private defense attorneys could be somehow forbidden from representing guilty clients. Hence, the needed reforms to the system must come from individual attorneys committing to work for the right reasons. For those who strive to ensure citizens' rights, or who truly believe their clients are innocent, providing a defense is a noble undertaking. But for those whose overriding motivation is greed, that legally "zealous defense" is ethically indefensible.

*When Defense Is Indefensible* ©UWorld

Annotations for this passage can be found in the CARS Passage Booklet.

## 3b. Passage Applications to New Context Question 3

A 1993 opinion of the D.C. Bar stated that in certain situations, a defense attorney who discovers their client intends to lie on the stand cannot choose to withdraw from the case. How is this opinion most relevant to the information in the passage?

- A. It shows that an attorney's motivations may be ethically indefensible.
- B. It exemplifies a tension between attorneys' professional obligations and ethics in general.
- C. It explains why reforms to the system must come from individual attorneys.
- D. It illustrates the effect of wealth and class disparities on defense attorney ethics.

See next page for the *strategy-based explanation* of this question.

Chapter 7: Skill 3 – Reasoning Beyond the Text

> **3b. Passage Applications to New Context Question 3**
> **Strategy-based Explanation**
>
> A 1993 opinion of the D.C. Bar stated that in certain situations, a defense attorney who discovers their client intends to lie on the stand cannot choose to withdraw from the case. How is this opinion most relevant to the information in the passage?
>
> A. It shows that an attorney's motivations may be ethically indefensible.
> B. It exemplifies a tension between attorneys' professional obligations and ethics in general.
> C. It explains why reforms to the system must come from individual attorneys.
> D. It illustrates the effect of wealth and class disparities on defense attorney ethics.

The question implies that defense attorneys may sometimes wish to withdraw from a case because of their clients' behavior, yet they are not allowed to do so. Thus, to determine the connection between this scenario and passage information, we need to consider what the passage says about defense attorneys' motivations and any restrictions they may face on their conduct when defending a client.

## Applying the Method

| Passage Excerpt | Step 1: Determine the Point of Connection |
|---|---|
| [P2] Why is it not shocking, then, that we tolerate the mirror image of this behavior from defense attorneys? For they engage in the same outrageous conduct, only on the other side. **Paid handsomely to represent even the vilest of clients**, they apply their oratorical prowess to manipulating jury perception, **keeping the guilty free and unpunished in exchange for money and status**. To the extent that this behavior takes place, **are some defense attorneys as unethical** as our hypothetical prosecutor? | The author implies that some defense attorneys are **unethical** because they **choose to defend guilty clients**. |
| [P3] It is worth distinguishing **two senses of the word "ethical"** here. For there are **standards of *professional* ethics** to which any attorney must conform, including standards particular to the defense. Most relevant to our purposes, a lawyer is obligated to provide their client with a "zealous defense." In other words, once an attorney takes on a client, they are **ethically bound to promote that client's rights, interests, or innocence**—in fact, *not* to do so would be *unethical*. Thus, one might try to suggest that this obligation undermines the claim that some defense attorneys act unethically. | However, the author also notes that attorneys have a **professional obligation** to promote their client's rights, interests, or innocence. |

**[P4]** However, **meeting that professional standard is not the same as being ethical in the general sense of the word**. The standard depends on the condition: *once an attorney takes on a client*. With the exception of court-appointed attorneys or public defenders, who are assigned to provide representation to those who would otherwise lack it, an attorney is never required to represent a defendant. Therefore, meeting one's obligations as a defense attorney does not necessarily make one ethical, because the choice to accept a specific case (and thus to incur those obligations in the first place) may itself be an unethical act.

Thus, the author draws a distinction between two types of ethics: **upholding professional ethics** is not the same as **being ethical in a general sense**.

In the same way, the question scenario contrasts what the attorney **personally thinks is right** (withdrawing from the case) and what the attorney is **professionally obligated to do** (continue defending the lying client).

Therefore, the point of connection between the question scenario and the passage is **the difference between ethics in general and defense attorneys' professional ethics**.

### Step 2: Apply That Connection to the Answer Choices

We found that the point of connection between the question scenario and the passage is **the difference between ethics in general and defense attorneys' professional ethics**. What is right or wrong in a general sense may differ from what is ethically required of an attorney when defending a client.

Therefore, the correct answer is **Choice B**, *it exemplifies a tension between attorneys' professional obligations and ethics in general*. Although it is wrong for a client to lie on the stand, a defense attorney may still have an ethical obligation to defend that client, leaving them unable to withdraw from the case.

For an alternative method of explanation based more specifically on passage evidence, you can view this question in the UWorld Qbank (sold separately).

Lesson 7.3

# RBT Subskill 3c. New Claim Support or Challenge

| Skill 3: Reasoning Beyond the Text ||
| --- | --- |
| **Subskill** | **Student Objective** |
| 3a. Exemplar Scenario for Passage Claims | Identify which scenario most exemplifies or logically follows from passage claims |
| 3b. Passage Applications to New Context | Apply passage information to a new situation or context |
| **3c. New Claim Support or Challenge** | **Evaluate how a new claim supports or challenges passage information** |
| 3d. External Scenario Support or Challenge | Determine which external scenario supports or challenges passage information |
| 3e. Applying Passage Perspectives | Apply the perspective of the author or another source in the passage to a new situation or claim |
| 3f. Additional Conclusions From New Information | Use new information to draw additional conclusions |
| 3g. Identifying Analogies | Identify analogies or similarities between passage ideas and ideas found outside the passage |

There are two RBT subskills that focus on how new information would affect passage claims or arguments. With the first of these subskills, **3c. New Claim Support or Challenge**, the new information is introduced in the question itself. The task is then to determine whether this new claim would strengthen, weaken, or (more rarely) not affect claims made in the passage. Some examples of these questions are:

- Suppose historians conclude that some parts of *The Art of War* believed to have been written by Sun Tzu were actually written by Sun Bin at a later date. If true, would this information support the author's argument?
- Assume research reveals that human beings have a tendency to make specific types of logical errors. How would this information affect the author's argument about the conflicting instructions in Aristotle's Rhetoric?
- According to the author, scholars agree that hunting would have been insufficient to provide for Jericho's sustenance. However, flint arrowheads have been discovered in the excavations of Jericho. Does this information challenge the scholars' view as it is described in the passage?

As you can see, the first two examples feature hypothetical claims while the third one presents its new claim as an established fact. However, this difference does not change how the questions are approached; in either case, the point is whether the new information, if it were true, would support, challenge, or leave unaffected claims made in the passage.

## Method for Answering 3c. New Claim Support or Challenge Questions

### Step 1: Restate the Relevant Claims or Argument

To determine the effect of the new claim, you must have a clear idea about what is being affected. Thus, the first step is to restate the passage claims or argument that are relevant to the topic. By breaking this information down into its components, you will be able to isolate the specific claim to which the new information most closely relates.

For instance, if you were considering the first question listed earlier, you would begin by considering what the author says about Sun Tzu's authorship of *The Art of War*, the writings of Sun Bin, and the relationship between the two. You would then have a framework to judge how the author's claims about these men and their writings would be affected by the hypothetical historical discovery.

### Step 2: Identify the New Claim's Connection

Using the framework you have established, the next step is to identify the specific passage information to which the new claim connects. Typically, the new claim and the passage information will have a common element through which they either reinforce each other or stand in contrast.

To continue our previous example, suppose the passage author presented an argument about when *The Art of War* was written, or one about the similarities between the two men's writings. Either topic would be impacted by a discovery that Sun Bin wrote material traditionally attributed to Sun Tzu. Thus, you would likely recognize a point of commonality between the author's argument and the new information introduced in the question.

### Step 3: Determine the New Claim's Effect

As in the examples given, some questions will specify that you are looking for supporting or challenging information, while others will just ask what effect the claim has. Either way, your comparison between the new claim and the original passage information should make their relationship clear. The new claim will either strengthen, weaken, or (again, more rarely) be irrelevant to the claims made in the passage. In all but those rare cases, when answering these questions, you should be able to indicate the passage information that the new claim supports or challenges.

## Subskill 3c. New Claim Support or Challenge
### (4 Practice Questions)

We now look at four examples from the UWorld Qbank. In each case, the passage and question are first given without commentary, allowing you to practice applying the method yourself. Then, the passage and question are presented again, this time with annotations of the passage (see the CARS Passage Booklet) and a step-by-step explanation of the question using the described method.

# Passage K: Lengthening the School Day

There may be reasons to reject the idea of lengthening the school day. None of them, however, are *good* reasons. Rather, the supposed demerits of such a proposal fall easily in the face of its numerous financial and social benefits for families.

The greatest of these benefits lies in reducing the need for childcare. It is a curious fact of American life that the adult's work schedule and the child's school schedule are misaligned. Children rise with the sun to head to classes, only to be sent home again hours before parents return from their jobs. In a society where, more often than not, both parents work, this discordance creates the need for an expensive arrangement to fill the gap in families' routines. For instance, studies show that in 2016, childcare costs accounted for 9.5 to 17.5 percent of median family income, depending on the state. Today, 40 percent of families nationwide spend over 15 percent of their income on childcare. Transportation to and from care sites only adds to that expense.

An additional advantage of an extended school day would be to allow for greater diversity and depth in curricula. Schools across the country have increasingly cut instruction in arts, music, and physical education (as well as recess) in order to meet objectives in reading and math. While this unfortunate state of affairs can be partially blamed on overzealous attention to standardized tests, it points to the larger deleterious trend of narrowing students' instruction. With a longer school day, such eliminated subjects can be restored, enriching students with a more well-rounded education.

To this proposal, however, critics may object that the added time would impose strain on educators. Can we truly ask schoolteachers—already among the most overworked individuals in society—to endure even more hours in the classroom? The answer is that a lengthened school day need not distress teachers nor add to their already cumbersome workload. By providing for additional areas of study in the arts and humanities, the extension would give schools cause to hire new, perhaps specialized, faculty to offer these courses. Moreover, the time could also be allocated to sports, academic clubs, and other extracurricular activities.

However, this point speaks to another objection, namely, the cost of adjusting the school day. Whether through paying current teachers more or hiring new ones, implementing such a proposal would entail a significant financial expenditure. There are at least two responses to this line of thinking. First is that this increase in the cost of schooling would be offset and likely surpassed by the aforementioned savings in childcare. Thus, while it is true that schools would require greater funding (likely necessitating higher property taxes), parents would ultimately pay the same or less overall, with greater educational opportunities for their children and fewer transportational burdens. Second is that schools should be better funded regardless. Recently, some schools—especially those in rural areas—have even reduced school weeks to only four days as a cost-saving measure. It is beyond dispute that schools across the board both need and deserve a radically increased investment from citizens. Lengthening the school day is simply one manifestation of how such funding should be utilized.

With this one change, states can coordinate the lives of parents and children, reduce the need for costly childcare, and expand curricular offerings. These worthy and desirable aims provide a clear justification for extending the school day.

*Lengthening the School Day* ©UWorld

Annotations for this passage can be found in the CARS Passage Booklet.

## 3c. New Claim Support or Challenge Question 1

Suppose a politician remarked that lengthening the school day would essentially make schools childcare facilities. What effect would this statement have on claims made in the passage?

    A.  It would support the author's claim that a longer school day would benefit parents.

    B.  It would reinforce the arguments for keeping the school day as it is.

    C.  It would challenge the author's claim that transportation to childcare sites burdens parents.

    D.  It would have no effect on the claims made in the passage.

See next page for the *strategy-based explanation* of this question.

Chapter 7: Skill 3 – Reasoning Beyond the Text

---

### 3c. New Claim Support or Challenge Question 1
### Strategy-based Explanation

Suppose a politician remarked that lengthening the school day would essentially make schools childcare facilities. What effect would this statement have on claims made in the passage?

A. It would support the author's claim that a longer school day would benefit parents.
B. It would reinforce the arguments for keeping the school day as it is.
C. It would challenge the author's claim that transportation to childcare sites burdens parents.
D. It would have no effect on the claims made in the passage.

---

The question refers to "claims made in the passage" without providing any details about those claims. However, we know the passage overall is an argument for lengthening the school day. Accordingly, we can think of this question as asking: "How does the politician's statement support or challenge the argument that the school day should be lengthened?" In addition, since the politician's statement relates specifically to childcare, we can focus on the claims in the argument that deal with that topic.

## Applying the Method

| Passage Excerpt | Step 1: Restate the Relevant Claims or Argument |
|---|---|
| **[P1]** There may be reasons to reject the idea of lengthening the school day. None of them, however, are *good* reasons. Rather, the supposed demerits of such a proposal fall easily in the face of its numerous financial and social benefits for families. | The author connects a longer school day with the issue of childcare through the following claims: |
| **[P2] The greatest of these benefits lies in reducing the need for childcare.** It is a curious fact of American life that **the adult's work schedule and the child's school schedule are misaligned.** Children rise with the sun to head to classes, only to be **sent home again hours before parents return from their jobs.** In a society where, more often than not, both parents work, this discordance **creates the need for an expensive arrangement to fill the gap** in families' routines. For instance, studies show that in 2016, childcare costs accounted for 9.5 to 17.5 percent of median family income, depending on the state. Today, 40 percent of families nationwide spend over 15 percent of their income on childcare. | **1.** Afterschool childcare is currently needed because **the school day ends hours before parents return from their jobs**. This mismatch forces parents to depend on **an expensive arrangement to fill that gap in time**, where children must leave school and be **sent to separate sites for childcare**. |

**Transportation to and from care sites** only adds to that expense.

**[P6]** With this one change, states can **coordinate the lives of parents and children, reduce the need for costly childcare**, and expand curricular offerings. These worthy and desirable aims provide a clear justification for extending the school day.

2. However, a lengthened school day would **coordinate parents' and students' schedules,** removing that gap in time.

3. Therefore, we should lengthen the school day to **reduce parents' need for costly childcare**.

### Step 2: Identify the New Claim's Connection

In the question scenario, the politician claims that lengthening the school day would essentially make schools childcare facilities.

If this claim is true, it means that lengthening the school day would keep children at school instead of at a separate childcare site.

Therefore, the politician's claim connects to the author's claims about the **gap in time between the end of the school day and the end of the work day** and the resulting need for **transportation to and from care sites**. If children were kept at school until parents returned from their jobs, **there would be no gap in time, eliminating the need for a separate childcare arrangement**.

### Step 3: Determine the New Claim's Effect

We determined that if the politician's claim is true, then **there would be no gap in time** between children's and parents' schedules, **eliminating the need for a separate childcare arrangement**. Thinking back to the passage argument, this is exactly what the author advocates: lengthening the school day in order to "**coordinate the lives of parents and children, [and] reduce the need for costly childcare**." Therefore, the effect of the politician's statement would be to **support the author's claims** that a lengthened school day would help to solve the problem of afterschool childcare.

Turning to the answer choices, the only one that refers to supporting the author's claims is **Choice A**, *It would support the author's claim that a longer school day would benefit parents*. Specifically, parents would benefit from not having to send kids to an expensive, separate childcare site after school. Thus, **Choice A** is the correct answer.

For an alternative method of explanation based more specifically on passage evidence, you can view this question in the UWorld Qbank (sold separately).

## Passage H: For Whom the Bell Toils

In nineteenth-century America, most people dismissed the notion that someone might assassinate the president. The presumption was based not only on ethics but practicality: a president's term is inherently limited, and an unpopular one could be voted out of office. Therefore, it was reasoned, there would be no need to consider removal through violence. This belief persisted even after the shocking murder of Abraham Lincoln in 1865, which was viewed as an aberration. Thus it was that on July 2, 1881, Charles Guiteau could simply walk up to President James A. Garfield and shoot him in broad daylight. As Richard Menke portrays events, "Guiteau was in fact a madman who had come to identify with a disgruntled wing of the Republican Party after his deranged fantasies of winning a post from the new administration had come to nothing." Believing that God had told him to kill the president, Guiteau thought this act would garner fame for his religious ideas and thereby help to usher in the Apocalypse.

In an interesting parallel, Garfield had felt a sense of divine purpose for his own life after surviving a near-drowning as a young man. Unlike Guiteau's fanatical ravings, however, Garfield's vision worked to the betterment of himself and the world. Candice Millard describes his ascent from extreme poverty to incredible excellence in college, where "by his second year…they made him a professor of literature, mathematics, and ancient languages." Garfield would go on to join the Union Army, where he attained the rank of major general and argued that black soldiers should receive the same pay as their white compatriots. While serving in the war he was nominated for the House of Representatives but accepted the seat only after President Lincoln declared that the country had more need of him as a congressman than as a general. The reluctant politician would later himself become president under similar circumstances, after multiple factions of a deadlocked Republican convention unexpectedly nominated him instead of their original candidates in 1880. An honest man who opposed corruption within the party, Garfield strove both to heal the fractures of the Civil War and to uphold the aims for which it was fought, until "the equal sunlight of liberty shall shine upon every man, black or white, in the Union."

Although Guiteau's bullet would ultimately dim this light for Garfield, the president actually survived the initial attack and for a time appeared headed for recovery. Tragically, however, the hubris shown by his main physician, Dr. Willard Bliss, would lead instead to weeks of prolonged suffering. None of the doctors who examined Garfield were able to locate the bullet, and its lingering presence—along with the unwashed hands of the doctors who probed for it—led to an infection. As the president's condition worsened, inventor Alexander Graham Bell attempted to adapt his patented telephone technology to locate foreign metal in the human body. Inspired by speculation that the bullet's electromagnetic properties might be detectable, Bell used his newly developed "Induction Balance" device to listen for the sounds of electrical interference he hoped would isolate the site of the bullet.

Unfortunately, Bell's searches were unsuccessful. Like Garfield's doctors, he had been looking for the bullet in the wrong area. Menke asserts that Bell's efforts "would probably have fallen short" regardless. However, other historians suggest that Dr. Bliss, unwilling to consider challenges to his original assessment, prevented Bell from more thoroughly searching the president's body. Certainly, Bliss ignored the advice and protestations of other physicians, even as Garfield continued to decline. With death imminent, Garfield asked to be taken to his seaside cottage, where he died on the 19th of September.

*For Whom the Bell Toils* ©UWorld

Annotations for this passage can be found in the CARS Passage Booklet.

### 3c. New Claim Support or Challenge Question 2

The Republican convention of 1880 required 36 separate rounds of voting over the course of three days before the delegates were finally able to agree on a presidential candidate. This information most strongly supports which of the following passage claims?

   A. Some members of the Republican Party were engaged in corrupt politics.
   B. Garfield's nomination was a compromise to resolve a deadlock.
   C. Some Republicans were disgruntled with the state of the party.
   D. America still suffered from the lingering divisions of the Civil War.

See next page for the *strategy-based explanation* of this question.

## 3c. New Claim Support or Challenge Question 2
### Strategy-based Explanation

The Republican convention of 1880 required 36 separate rounds of voting over the course of three days before the delegates were finally able to agree on a presidential candidate. This information most strongly supports which of the following passage claims?

A. Some members of the Republican Party were engaged in corrupt politics.
B. Garfield's nomination was a compromise to resolve a deadlock.
C. Some Republicans were disgruntled with the state of the party.
D. America still suffered from the lingering divisions of the Civil War.

The question refers to the Republican convention of 1880, a topic that is discussed only briefly in the passage. Therefore, we can quickly focus our attention where that discussion appears, near the end of Paragraph 2.

## Applying the Method

*Passage Excerpt*

**[P2]** While serving in the war he was nominated for the House of Representatives but accepted the seat only after President Lincoln declared that the country had more need of him as a congressman than as a general. The reluctant politician would later himself become president under similar circumstances, after multiple factions of a deadlocked Republican convention unexpectedly nominated him instead of their original candidates in 1880. An honest man who opposed corruption within the party, Garfield strove both to heal the fractures of the Civil War and to uphold the aims for which it was fought, until "the equal sunlight of liberty shall shine upon every man, black or white, in the Union."

### Step 1: Restate the Relevant Claims or Argument

The passage contains only one sentence about the 1880 Republican convention. It tells us that Garfield's nomination for president came when multiple factions of a deadlocked Republican convention unexpectedly nominated him instead of their original candidates in 1880.

### Step 2: Identify the New Claim's Connection

Given how little the author says about the 1880 Republican convention, the new claim's connection is straightforward. The fact that the convention required 36 separate rounds of voting over the course of three days to pick a candidate clearly relates to the **deadlock between multiple factions** that the author describes.

### Step 3: Determine the New Claim's Effect

Based on the small amount of information that the passage contains about the 1880 Republican convention, we saw that the claim being asked about relates to the described **deadlock between multiple factions**. This new information about just how long the nominating process took serves to reinforce and elaborate on those factions' inability to agree on a candidate.

Accordingly, the correct answer is **Choice B**, *Garfield's nomination was a compromise to resolve a deadlock*. Since it took 36 rounds of voting over three days for a candidate to be selected, it is easy to see why such a compromise was needed.

---

### Beyond the Text...But Not Way Beyond

As you looked at the answer choices, some of the others might also have seemed plausible. For instance, Choice C talks about Republicans being "disgruntled with the state of the party" and Choice D mentions "lingering divisions," both of which could help explain why the nomination process took so long.

Notice, however, that each of those answers involves an extra step of reasoning that the correct answer does not. The claim being asked about supports the idea that the convention was deadlocked, regardless of *why* that deadlock may have existed. Thus, while it could be true that some Republicans were disgruntled or the country had lingering divisions, the deadlock existed either way, making **Choice B** the superior answer.

As a general rule, even when you're reasoning beyond the text, **the correct answer is more likely to be the one that requires fewer steps to reach**. Keeping this principle in mind can be helpful when answering many CARS questions.

---

For an alternative method of explanation based more specifically on passage evidence, you can view this question in the UWorld Qbank (sold separately).

## Passage L: When Defense Is Indefensible

Suppose a prosecutor is considering whether to bring a case to trial. He is not sure that the suspect is guilty—in fact, based on the evidence, it's more likely that the suspect is *not* guilty. Nevertheless, he feels confident he can secure a guilty verdict. His powers of persuasion are considerable, and there's a good chance he could trick a jury into believing the evidence is strong instead of weak. In addition, the case is high profile and could be very lucrative; winning would likely lead to a substantial raise or promotion. He decides to charge the suspect, and ultimately succeeds in persuading the jury to convict.

Looking at this situation, most of us would easily judge the prosecutor as extremely unethical. His conduct is outrageous and wrong—he clearly acted with corrupt intent, perpetrating injustice in order to profit financially. Why is it not shocking, then, that we tolerate the mirror image of this behavior from defense attorneys? For they engage in the same outrageous conduct, only on the other side. Paid handsomely to represent even the vilest of clients, they apply their oratorical prowess to manipulating jury perception, keeping the guilty free and unpunished in exchange for money and status. To the extent that this behavior takes place, are some defense attorneys as unethical as our hypothetical prosecutor?

It is worth distinguishing two senses of the word "ethical" here. For there are standards of *professional* ethics to which any attorney must conform, including standards particular to the defense. Most relevant to our purposes, a lawyer is obligated to provide their client with a "zealous defense." In other words, once an attorney takes on a client, they are ethically bound to promote that client's rights, interests, or innocence—in fact, *not* to do so would be *unethical*. Thus, one might try to suggest that this obligation undermines the claim that some defense attorneys act unethically.

However, meeting that professional standard is not the same as being ethical in the general sense of the word. The standard depends on the condition: *once an attorney takes on a client*. With the exception of court-appointed attorneys or public defenders, who are assigned to provide representation to those who would otherwise lack it, an attorney is never required to represent a defendant. Therefore, meeting one's obligations as a defense attorney does not necessarily make one ethical, because the choice to accept a specific case (and thus to incur those obligations in the first place) may itself be an unethical act.

Moreover, the role of court-appointed attorneys is to help protect the rights of citizens who cannot secure their own representation, usually for financial reasons. Although preserving those rights is necessary to uphold justice, this situation highlights how wealth and class contribute to *injustice*. While some defendants possess the means to hire top-level private lawyers, others must depend on public servants—frequently less experienced lawyers from overloaded, understaffed agencies. As a result, the rich are more likely to escape conviction even when they are guilty, and the poor are more likely to be convicted even when they are innocent.

It is doubtful that private defense attorneys could be somehow forbidden from representing guilty clients. Hence, the needed reforms to the system must come from individual attorneys committing to work for the right reasons. For those who strive to ensure citizens' rights, or who truly believe their clients are innocent, providing a defense is a noble undertaking. But for those whose overriding motivation is greed, that legally "zealous defense" is ethically indefensible.

*When Defense Is Indefensible* ©UWorld

Annotations for this passage can be found in the CARS Passage Booklet.

## 3c. New Claim Support or Challenge Question 3

The author argues that it is unethical for defense attorneys to take certain defendants as clients. However, the Sixth Amendment guarantees every defendant the right to legal representation. Can the author's argument accommodate this fact?

- A. Yes, because the needed reforms to the legal system could include the Sixth Amendment.
- B. Yes, because the author's ethical claim applies specifically to private attorneys.
- C. No, because if all attorneys accepted the author's argument, then some defendants could not get representation.
- D. No, because a collective refusal to represent certain defendants could lead to innocent people being convicted.

See next page for the *strategy-based explanation* of this question.

Chapter 7: Skill 3 – Reasoning Beyond the Text

> ### 3c. New Claim Support or Challenge Question 3
> ### Strategy-based Explanation
>
> The author argues that it is unethical for defense attorneys to take certain defendants as clients. However, the Sixth Amendment guarantees every defendant the right to legal representation. Can the author's argument accommodate this fact?
>
> A. Yes, because the needed reforms to the legal system could include the Sixth Amendment.
> B. Yes, because the author's ethical claim applies specifically to private attorneys.
> C. No, because if all attorneys accepted the author's argument, then some defendants could not get representation.
> D. No, because a collective refusal to represent certain defendants could lead to innocent people being convicted.

The question asks whether the author's argument "can accommodate" the Sixth Amendment's guarantee of legal representation, or, in other words, whether the argument can be accepted despite the challenge this new information appears to pose. Accordingly, we need to consider the details of that argument to determine whether it conflicts with the Sixth Amendment as the question suggests it might.

## Applying the Method

*Passage Excerpt*

**Step 1: Restate the Relevant Claims or Argument**

[P2] Looking at this situation, most of us would easily judge the prosecutor as extremely unethical. His conduct is outrageous and wrong—he clearly acted with corrupt intent, perpetrating injustice in order to profit financially. Why is it not shocking, then, that we tolerate the mirror image of this behavior from defense attorneys? For they engage in the same outrageous conduct, only on the other side. **Paid handsomely to represent even the vilest of clients,** they apply their oratorical prowess to manipulating jury perception, **keeping the guilty free and unpunished in exchange for money and status.** To the extent that this behavior takes place, are some defense attorneys as unethical as our hypothetical prosecutor?

The author's argument is developed through several parts of the passage.

Paragraph 2 calls it unethical for defense attorneys to **keep the guilty free and unpunished in exchange for money and status**.

[P4] However, meeting that professional standard is not the same as being ethical in the general sense of the word. The standard depends on the condition: *once an attorney takes on a client.* **With the exception of court-appointed attorneys or public defenders, who are assigned to provide**

Paragraphs 4 and 5 distinguish between public and private attorneys. **Public attorneys are**

290

Chapter 7: Skill 3 – Reasoning Beyond the Text

representation to those who would otherwise lack it, an attorney is never required to represent a defendant. Therefore, meeting one's obligations as a defense attorney does not necessarily make one ethical, because the choice to accept a specific case (and thus to incur those obligations in the first place) may itself be an unethical act.

[P5] Moreover, the role of court-appointed attorneys is to help protect the rights of citizens who cannot secure their own representation, usually for financial reasons. Although preserving those rights is necessary to uphold justice, this situation highlights how wealth and class contribute to *injustice*.

[P6] It is doubtful that private defense attorneys could be somehow forbidden from representing guilty clients. Hence, the needed reforms to the system must come from individual attorneys committing to work for the right reasons. For those who strive to ensure citizens' rights, or who truly believe their clients are innocent, providing a defense is a noble undertaking. But for those whose overriding motivation is greed, that legally "zealous defense" is ethically indefensible.

assigned to defend people to ensure their representation. By contrast, private attorneys can choose—perhaps unethically—whom they want to represent.

Finally, Paragraph 6 calls on private attorneys to work for the right reasons instead of acting from an overriding motivation of greed.

### Step 2: Identify the New Claim's Connection

The question introduced the new claim that the Sixth Amendment guarantees every defendant the right to legal representation.

However, in reviewing the author's argument, we can see that it also refers to guaranteed legal representation. Paragraphs 4 and 5 discuss how public attorneys must defend anyone to whom they are assigned, protecting the rights of citizens who would otherwise lack representation. By contrast, private attorneys can refuse to accept certain clients if they wish.

Therefore, the Sixth Amendment's guarantee of legal representation connects most to this information about the obligations of public attorneys and their difference from private attorneys.

## Step 3: Determine the New Claim's Effect

The Sixth Amendment guarantees all defendants legal representation, and as we have seen, the author's argument also refers to such a guarantee: while private attorneys choose their clients, **public attorneys are obligated to ensure representation for whomever they are assigned to defend**. Furthermore, although the passage states that private attorneys can choose—perhaps unethically—whom they want to represent, the author does not make a similar claim about the ethics of public attorneys. Accordingly, the author's argument would seem to be **consistent with the Sixth Amendment's guarantee** rather than in conflict with it.

Therefore, the correct answer is **Choice B**, *Yes, because the author's ethical claim applies specifically to private attorneys*. This answer choice reflects the aspects of the author's argument that are relevant to the Sixth Amendment: while the author argues that private defense attorneys should refrain from defending the guilty, this claim does not apply to public defense attorneys, who would still serve to guarantee defendants' Sixth Amendment rights.

For an alternative method of explanation based more specifically on passage evidence, you can view this question in the UWorld Qbank (sold separately).

# Passage D: American Local Motives

Locomotives were invented in England, with the first major railroad connecting Liverpool and Manchester in 1830. However, it was in America that railroads would be put to the greatest use in the nineteenth century. On May 10, 1869, the Union Pacific and Central Pacific lines met at Promontory Point, Utah, joining from opposite directions to complete a years-long project—the Transcontinental Railroad. This momentous event connected the eastern half of the United States with its western frontier and facilitated the construction of additional lines in between. As a result, journeys that had previously taken several months by horse and carriage now required less than a week's travel. By 1887 there were nearly 164,000 miles of railroad tracks in America, and by 1916 that number had swelled to over 254,000.

While the United States still has the largest railroad network in the world, it operates largely in the background of American life, and citizens no longer view trains with the sense of importance those machines once commanded. Nevertheless, the economic and industrial advantages those citizens enjoy today would not have been possible without America's history of trains; as Tom Zoellner reminds us, "Under the skin of modernity lies a skeleton of railroad tracks." Although airplanes and automobiles have now assumed greater prominence, the time has arrived for the resurgence of railroads. A revitalized and advanced railway system would confer numerous essential benefits on both the United States and the globe.

The chief obstacles to garnering support for such a project are the current dominance of the automobile and the languishing technology of existing railroads. In a sense these two obstacles are one, as American dependence on personal automobiles is partially due to the paucity of rapid public transportation. The railroads of Europe and Japan, by comparison, have vastly outpaced their American counterparts. Japan has operated high-speed rail lines continuously since 1964, and in 2007, a French train set a record of 357 miles per hour. While that speed was achieved under tightly controlled conditions, it still speaks to the great disparity in railroad development between the United States and other countries since the mid-twentieth century. British trains travel at speeds much higher than those in America, where both the trains themselves and the infrastructure to support them have simply been allowed to fall behind. In much of Europe it is common for trains to travel at close to 200 miles per hour.

To invest in a modern network of railroads would improve the United States in much the same way that the first railroads did in the nineteenth and early twentieth centuries. A high-speed passenger rail system would dramatically transform American life as travel between cities and states became quicker and more convenient, encouraging commerce, business, and tourism. Such a system would also make important strides in environmental preservation. According to a 2007 British study, "CO2 emissions from aircraft operations are...at least five times greater" than those from high-speed trains. For similar reasons, Osaka, Japan, was ranked as "the best…green transportation city in Asia" by the 2011 *Green City Index*. As Lee-in Chen Chiu notes in *The Kyoto Economic Review*, Osakans travel by railway more than twice as much as they travel by car.

It is true that developing a countrywide high-speed rail system would come with significant costs. However, that was also true of the original Transcontinental Railroad, as indeed it is with virtually any great project undertaken for the public good. We should thus move ahead with confidence that the rewards will outweigh the expenditure as citizens increasingly choose to travel by train. Both for society's gain and the crucial well-being of the planet, our path forward should proceed upon rails.

*American Local Motives* ©UWorld

Annotations for this passage can be found in the CARS Passage Booklet.

### 3c. New Claim Support or Challenge Question 4

How would the author's argument be impacted if studies suggest there is an increasing trend of people using ride-sharing services for transportation?

- A. It would refute the author's suggestion that trains would help fill a transportational need.
- B. It would weaken the author's comparison between the Transcontinental Railroad and high-speed rail.
- C. It would support the author's assumption that the public could be willing to depend less on vehicle ownership.
- D. It would reinforce the author's claim that automobiles have made public transportation less important.

See next page for the *strategy-based explanation* of this question.

> ### 3c. New Claim Support or Challenge Question 4
> ### Strategy-based Explanation
>
> How would the author's argument be impacted if studies suggest there is an increasing trend of people using ride-sharing services for transportation?
>
> A. It would refute the author's suggestion that trains would help fill a transportational need.
> B. It would weaken the author's comparison between the Transcontinental Railroad and high-speed rail.
> C. It would support the author's assumption that the public could be willing to depend less on vehicle ownership.
> D. It would reinforce the author's claim that automobiles have made public transportation less important.

In arguing that high-speed rail should be built in America, the author assumes that people would use that form of transportation if it were available. Accordingly, we can think of the question as asking: how would that assumption be impacted by a current trend of people using ride-sharing services?

## Applying the Method

*Passage Excerpt*

[P3] The chief obstacles to garnering support for such a project are the current dominance of the automobile and the languishing technology of existing railroads. In a sense these two obstacles are one, as **American dependence on personal automobiles is partially due to the paucity of rapid public transportation**. The railroads of Europe and Japan, by comparison, have vastly outpaced their American counterparts. Japan has operated high-speed rail lines continuously since 1964, and in 2007, a French train set a record of 357 miles per hour. While that speed was achieved under tightly controlled conditions, it still speaks to the great disparity in railroad development between the United States and other countries since the mid-twentieth century. British trains travel at speeds much higher than those in **America, where both the trains themselves and the infrastructure to support them have simply been allowed to fall behind**. In much of Europe it is common for trains to travel at close to 200 miles per hour.

### Step 1: Restate the Relevant Claims or Argument

Part of the author's argument involves explaining why rail travel is not a bigger part of modern American life. According to the author:

1. The current **American dependence on personal automobiles** can be explained by the **paucity of rapid public transportation**.

2. Similarly, rail travel is not more prominent because **both the trains themselves and the infrastructure to support them have simply been allowed to fall behind**.

**[P4]** To invest in a modern network of railroads would improve the United States in much the same way that the first railroads did in the nineteenth and early twentieth centuries.  **A high-speed passenger rail system would dramatically transform American life** as travel between cities and states became quicker and more convenient, encouraging commerce, business, and tourism.  Such a system would also make important strides in environmental preservation.

**[P5]** We should thus move ahead with confidence that **the rewards will outweigh the expenditure as citizens increasingly choose to travel by train**.  Both for society's gain and the crucial well-being of the planet, our path forward should proceed upon rails.

3. However, if citizens did have access to a **high-speed passenger rail system, it would dramatically transform American life** for the better.

4. Therefore, we should build high-speed rail because **the rewards will outweigh the expenditure as citizens increasingly choose to travel by train.**

In summary, the author argues that people depend on personal cars partly because current trains are sub-par.  But, if we had modern high-speed trains, people would choose that form of travel.

### Step 2: Identify the New Claim's Connection

If there is an increasing trend of people using ride-sharing services, then that means people are becoming less dependent on vehicle ownership for travel.

Thus, the new claim connects to the author's assertion about **dependence on personal automobiles**.  Given the trend described, people are **interested in alternatives to that dependence**.

### Step 3: Determine the New Claim's Effect

As we have seen, the trend described in the question would suggest that people are **interested in alternatives to dependence on personal automobiles**. Accordingly, the impact of such a trend on the author's argument is best reflected by **Choice C**, *It would support the author's assumption that the public could be willing to depend less on vehicle ownership*. If people are increasingly using ride-sharing services, then they are willing to use transportation methods that do not require them to own a car.

Choice D, *It would reinforce the author's claim that automobiles have made public transportation less important*, might also have seemed appealing. However, when we look at this answer choice more closely, we can see that the author does not actually claim that automobiles made public transportation less important. Rather, the author claims that people depend on automobiles because effective public transportation is *unavailable*. By contrast, the author does assume people would be willing to depend less on vehicle ownership, as stated in **Choice C**.

For an alternative method of explanation based more specifically on passage evidence, you can view this question in the UWorld Qbank (sold separately).

Lesson 7.4

# RWT Subskill 3d. External Scenario Support or Challenge

| Skill 3: Reasoning Beyond the Text ||
|---|---|
| **Subskill** | **Student Objective** |
| 3a. Exemplar Scenario for Passage Claims | Identify which scenario most exemplifies or logically follows from passage claims |
| 3b. Passage Applications to New Context | Apply passage information to a new situation or context |
| 3c. New Claim Support or Challenge | Evaluate how a new claim supports or challenges passage information |
| **3d. External Scenario Support or Challenge** | **Determine which external scenario supports or challenges passage information** |
| 3e. Applying Passage Perspectives | Apply the perspective of the author or another source in the passage to a new situation or claim |
| 3f. Additional Conclusions From New Information | Use new information to draw additional conclusions |
| 3g. Identifying Analogies | Identify analogies or similarities between passage ideas and ideas found outside the passage |

The second RBT subskill that focuses on how new information would affect passage claims or arguments is **3d. External Scenario Support or Challenge**. While similar to **3c. New Claim Support or Challenge**, there is a key difference between the two. Whereas **3c** questions provide new information and ask about its effect, **3d** questions do the reverse: they describe an effect on a claim or argument, then ask which of the listed scenarios would produce it.

These questions vary in the amount of detail provided. For instance, one question might be fairly specific, such as:

- Which of the following, if true, would most *challenge* Marina Warner's analysis of the events in *Not Now, Bernard* (Paragraph 5)?

while another question might be more general, like:

- Which of the following discoveries would most support the author's argument?

The method for approaching these questions is the same in either case, but the more general questions may take additional mental effort since you are starting with less information.

# Method for Answering 3d. External Scenario Support or Challenge Questions

### Step 1: Restate the Relevant Claims or Argument

As with 3c questions, the first step is to restate the relevant passage claims or argument to break this information down into its components. Reorganizing the passage claims in your mind can help make it clearer what would support or challenge them.

For example, in approaching the first question listed, you would restate what Marina Warrner says about *Not Now, Bernard* in a simpler, more organized form. This process would then make the relevant passage claims much easier to analyze.

### Step 2: Analyze the Basis of Support

After restating the relevant passage claims, the next step is to analyze the underlying points that those claims are based on. Any scenario that would make those points more likely to be true would thus also support the passage claims. Similarly, any scenario that would make those points less likely to be true would challenge the passage claims.

In our previous example, suppose you determined that Marina Warner makes a particular assumption that motivates her analysis of *Not Now, Bernard*. In that case, any information which would cast doubt on that assumption would also challenge her analysis. Therefore, you would know that such information would constitute a possible answer to the question.

### Step 3: Compare Answer Choices with That Basis

As you turn to the answer choices, you will see that one relates to the basis of support you identified in Step 2. You should then be able to indicate the specific passage information that this new scenario supports or challenges.

Chapter 7: Skill 3 – Reasoning Beyond the Text

## Subskill 3d. External Scenario Support or Challenge
### (5 Practice Questions)

We now look at three examples from the UWorld Qbank. In each case, the passage and question are first given without commentary, allowing you to practice applying the method yourself. Then, the passage and question are presented again, this time with annotations of the passage (see the CARS Passage Booklet) and a step-by-step explanation of the question using the described method.

## Passage D: American Local Motives

Locomotives were invented in England, with the first major railroad connecting Liverpool and Manchester in 1830. However, it was in America that railroads would be put to the greatest use in the nineteenth century. On May 10, 1869, the Union Pacific and Central Pacific lines met at Promontory Point, Utah, joining from opposite directions to complete a years-long project—the Transcontinental Railroad. This momentous event connected the eastern half of the United States with its western frontier and facilitated the construction of additional lines in between. As a result, journeys that had previously taken several months by horse and carriage now required less than a week's travel. By 1887 there were nearly 164,000 miles of railroad tracks in America, and by 1916 that number had swelled to over 254,000.

While the United States still has the largest railroad network in the world, it operates largely in the background of American life, and citizens no longer view trains with the sense of importance those machines once commanded. Nevertheless, the economic and industrial advantages those citizens enjoy today would not have been possible without America's history of trains; as Tom Zoellner reminds us, "Under the skin of modernity lies a skeleton of railroad tracks." Although airplanes and automobiles have now assumed greater prominence, the time has arrived for the resurgence of railroads. A revitalized and advanced railway system would confer numerous essential benefits on both the United States and the globe.

The chief obstacles to garnering support for such a project are the current dominance of the automobile and the languishing technology of existing railroads. In a sense these two obstacles are one, as American dependence on personal automobiles is partially due to the paucity of rapid public transportation. The railroads of Europe and Japan, by comparison, have vastly outpaced their American counterparts. Japan has operated high-speed rail lines continuously since 1964, and in 2007, a French train set a record of 357 miles per hour. While that speed was achieved under tightly controlled conditions, it still speaks to the great disparity in railroad development between the United States and other countries since the mid-twentieth century. British trains travel at speeds much higher than those in America, where both the trains themselves and the infrastructure to support them have simply been allowed to fall behind. In much of Europe it is common for trains to travel at close to 200 miles per hour.

To invest in a modern network of railroads would improve the United States in much the same way that the first railroads did in the nineteenth and early twentieth centuries. A high-speed passenger rail system would dramatically transform American life as travel between cities and states became quicker and more convenient, encouraging commerce, business, and tourism. Such a system would also make important strides in environmental preservation. According to a 2007 British study, "CO2 emissions from aircraft operations are...at least five times greater" than those from high-speed trains. For similar reasons, Osaka, Japan, was ranked as "the best...green transportation city in Asia" by the 2011 *Green City Index*. As Lee-in Chen Chiu notes in *The Kyoto Economic Review*, Osakans travel by railway more than twice as much as they travel by car.

It is true that developing a countrywide high-speed rail system would come with significant costs. However, that was also true of the original Transcontinental Railroad, as indeed it is with virtually any great project undertaken for the public good. We should thus move ahead with confidence that the rewards will outweigh the expenditure as citizens increasingly choose to travel by train. Both for society's gain and the crucial well-being of the planet, our path forward should proceed upon rails.

*American Local Motives* ©UWorld

Annotations for this passage can be found in the CARS Passage Booklet.

### 3d. External Scenario Support or Challenge Question 1

Which of the following statements, if true, would most *challenge* the author's argument for the value of building a high-speed rail system in America?

A. On average, the distance between travel destinations is shorter in Europe than it is in America.

B. The slower speeds of trains in America compared to those in Europe are due to safety issues with American trains.

C. The primary use of existing American trains is the transport of freight rather than passengers.

D. The process of constructing high-speed rail could affect the environment in unpredictable ways.

See next page for the *strategy-based explanation* of this question.

## 3d. External Scenario Support or Challenge Question 1
## Strategy-based Explanation

Which of the following statements, if true, would most *challenge* the author's argument for the value of building a high-speed rail system in America?

- A. On average, the distance between travel destinations is shorter in Europe than it is in America.
- B. The slower speeds of trains in America compared to those in Europe are due to safety issues with American trains.
- C. The primary use of existing American trains is the transport of freight rather than passengers.
- D. The process of constructing high-speed rail could affect the environment in unpredictable ways.

Since the question asks about challenging the author's argument, the correct answer will cast doubt on the reasons that argument is based on. Accordingly, answering this question is a matter of considering why the author claims a high-speed rail system would be valuable, then determining what information would count against those claims.

## Applying the Method

*Passage Excerpt*

**[P2]** Although airplanes and automobiles have now assumed greater prominence, the time has arrived for the resurgence of railroads. A revitalized and advanced railway system **would confer numerous essential benefits on both the United States and the globe**.

**[P4]** To invest in a modern network of railroads would improve the United States in much the same way that the first railroads did in the nineteenth and early twentieth centuries. A high-speed passenger rail system **would dramatically transform American life as travel between cities and states became quicker and more convenient, encouraging commerce, business, and tourism**. Such a system would **also make important strides in environmental preservation**. According to a 2007 British study, "$CO_2$ emissions from aircraft operations are...at least five times greater" than those from high-speed trains. For similar reasons, Osaka, Japan, was ranked as "the

### Step 1: Restate the Relevant Claims or Argument

The author argues that building a high-speed rail system in America would be valuable because:

1. Such a system would confer numerous essential benefits on both the United States and the globe.

2. Specifically, it would:
   - produce economic gains in America by encouraging commerce, business, and tourism; and
   - make important strides in environmental preservation because traveling by train is greener than traveling by plane or car.

best…green transportation city in Asia" by the 2011 *Green City Index*. As Lee-in Chen Chiu notes in *The Kyoto Economic Review*, Osakans travel by railway more than twice as much as they travel by car.

[P5] We should thus move ahead with confidence that the rewards will outweigh the expenditure as citizens increasingly choose to travel by train. Both for society's gain and the crucial well-being of the planet, our path forward should proceed upon rails.

## Step 2: Analyze the Basis of Support

The author's argument for building high-speed rail depends on the claims that it would cause both American economic gains and global environmental benefits. Therefore, challenging the author's argument means giving reason to believe one or both of the following:

1. High-speed rail **would not cause economic gains in America**.

2. High-speed rail **would not benefit the environment**.

Either of these statements, if true, would undermine the author's basis for arguing that building high-speed rail would be beneficial.

### Step 3: Compare Answer Choices with That Basis

As we have seen, challenging the basis for the author's argument would mean giving reason to think either that high-speed rail **would not cause economic gains in America** or that it **would not benefit the environment**.

Turning to the answer choices, we find that only one option would suggest either of those possibilities: **Choice D**, *The process of constructing high-speed rail could affect the environment in unpredictable ways*. If the effects on the environment would be unpredictable, then building high-speed rail might not help the environment and could even harm it. Therefore, this prospect would challenge the author's claim that building high-speed rail would **make important strides in environmental preservation**, showing **Choice D** to be correct.

For an alternative method of explanation based more specifically on passage evidence, you can view this question in the UWorld Qbank (sold separately).

## Passage G: Food Costs and Disease

Because frequent consumption of unhealthy foods is strongly linked with cardiometabolic diseases, one way for governments to combat those afflictions may be to modify the eating habits of the general public. Applying economic incentives or disincentives to various types of foods could potentially alter people's diets, leading to more positive health outcomes.

Utilizing national data from 2012 regarding food consumption, health, and economic status, Peñalvo et al. concluded that such price adjustments would help to prevent deaths related to cardiometabolic diseases. According to their analysis, increasing the prices of unhealthy foods such as processed meats and sugary sodas by 10%, while reducing the prices of healthy foods such as fruit and vegetables by 10%, would prevent an estimated 3.4% of yearly deaths in the U.S. Changing prices by 30% would have an even stronger effect, preventing an estimated 9.2% of yearly deaths. This data comports with that found in other countries, such as "previous modeling studies in South Africa and India, where a 20% SSB [sugar-sweetened beverage] tax was estimated to reduce diabetes prevalence by 4% over 20 years." The effects of price adjustments would be most pronounced on persons of lower socioeconomic status, as the researchers "found an overall 18.2% higher price-responsiveness for low versus high SES [socioeconomic status] groups."

This differential effect based on socioeconomic status contributes to concerns about such interventions, however. In *Harvard Public Health Review*, Kates and Hayward ask: "Well intentioned though they may be, at what point do these taxes overstep government influence on an individual's right to autonomy in decision-making? On whom does the increased financial burden of this taxation fall?" They note that taxes on sugar-sweetened beverages, for instance, "are likely to have a greater impact on low-income individuals…because individuals in those settings are more likely to be beholden to cost when making decisions about food."

However, "well intentioned though they may be," the worries that Kates and Hayward express are to some extent misguided. In particular, the idea that taxing unhealthy foods would burden those least able to afford it misses the point. Although the increased taxes would affect anyone who continued to purchase the items despite the higher prices, the goal of raising prices on unhealthy foods is precisely to dissuade people from buying them. As Kates and Hayward themselves remark, "Those in low-income environments may also be the largest consumers of obesogenic foods and therefore most likely to benefit from such a lifestyle change indirectly posed by SSB taxes." As the goal of the taxes is to promote those lifestyle changes, the financial burden objection is a non-starter.

Given this recognition, the question regarding autonomy constitutes a more substantial issue. Nevertheless, that concern also rests on a dubious assumption, as people's autonomy is not necessarily respected in the current situation either. The fact that those of lower socioeconomic status are more likely to have poorer diets suggests that such persons' food choices are the result of financial constraint, not fully autonomous, rational deliberation. Hence, by making healthy foods more affordable relative to unhealthy ones, government intervention might actually *facilitate* autonomous choices rather than hindering them.

On the other hand, suppose that the disproportionate consumption of unhealthy foods—and associated higher incidence of disease—among certain groups is not the result of financial hardship but rather the result of those persons' perceived self-interest. If so, that would suggest that members of these groups are being encouraged to persist in harmful dietary habits for the sake of corporate profits. In that case, violating autonomy for the sake of health may be permissible, as that practice would be morally preferable to the present system of corporate exploitation.

*Food Costs and Disease* ©UWorld

Annotations for this passage can be found in the CARS Passage Booklet.

### 3d. External Scenario Support or Challenge Question 2

Which of the following statements, if true, would most support the author's argument for price adjustments?

A. A study conducted in Mexico reported higher price responsiveness among low-income groups.
B. Many corporations are willing to decrease prices to avoid losing customers.
C. Kates and Hayward are funded by a sugar-sweetened beverage corporation.
D. Peñalvo et al. inferred people's socioeconomic status from their educational attainment.

See next page for the *strategy-based explanation* of this question.

Chapter 7: Skill 3 – Reasoning Beyond the Text

---

### 3d. External Scenario Support or Challenge Question 2
### Strategy-based Explanation

Which of the following statements, if true, would most support the author's argument for price adjustments?

A. A study conducted in Mexico reported higher price responsiveness among low-income groups.
B. Many corporations are willing to decrease prices to avoid losing customers.
C. Kates and Hayward are funded by a sugar-sweetened beverage corporation.
D. Peñalvo et al. inferred people's socioeconomic status from their educational attainment.

---

You probably recall that the passage first describes the potential benefits of price adjustments, then switches to addressing a concern about consumers' autonomy. Accordingly, the author's argument for price adjustments comes primarily in the first half of the passage, which discusses the effects of such adjustments in reducing deaths from disease.

## Applying the Method

*Passage Excerpt*

**[P1]** Because frequent consumption of unhealthy foods is strongly linked with cardiometabolic diseases, one way for governments to combat those afflictions may be to modify the eating habits of the general public. **Applying economic incentives or disincentives to various types of foods could potentially alter people's diets, leading to more positive health outcomes.**

**[P2]** Utilizing national data from 2012 regarding food consumption, health, and economic status, **Peñalvo et al. concluded that such price adjustments would help to prevent deaths related to cardiometabolic diseases**. According to their analysis, increasing the prices of unhealthy foods such as processed meats and sugary sodas by 10%, while reducing the prices of healthy foods such as fruit and vegetables by 10%, would prevent an estimated 3.4% of yearly deaths in the U.S. Changing prices by 30% would have an even stronger effect, preventing an estimated 9.2% of yearly deaths. **This data comports with that found in other countries**, such

### Step 1: Restate the Relevant Claims or Argument

Paragraph 1 implies that governments should:

- **adjust the prices of foods**, which would **alter people's diets**, which would **reduce cardiometabolic disease**.

Paragraph 2 then supports this proposal with data about the relationship between food prices and disease. For example:

- A study by Peñalvo et al. **correlates price adjustments and disease prevention in the U.S.**
- The author notes that **this data comports with that found in other countries**.
- The effects of the correlation are **most pronounced on persons of lower socioeconomic status**.

Accordingly, the author argues that **data supports** a proposal to **adjust food prices** in order to **combat cardiometabolic disease**, especially among low SES individuals.

as "previous modeling studies in South Africa and India, where a 20% SSB [sugar-sweetened beverage] tax was estimated to reduce diabetes prevalence by 4% over 20 years." **The effects of price adjustments would be most pronounced on persons of lower socioeconomic status**, as the researchers "found an overall 18.2% higher price-responsiveness for low versus high SES [socioeconomic status] groups."

### Step 2: Analyze the Basis of Support

The author's argument for price adjustments is based on a proposed **relationship between food prices and cardiometabolic disease** as well as **research data** that seems to establish that relationship.

Therefore, we can see that the author's argument would be supported by statements that:

- **confirm the correlation** between food prices and disease; or
- **reinforce the data** behind that relationship.

### Step 3: Compare Answer Choices with That Basis

We determined that the author's argument for price adjustments would be supported by statements that **confirm the correlation** between food prices and disease or **reinforce the data** behind that relationship. Accordingly, the answer choice that stands out is **Choice A**, *A study conducted in Mexico reported higher price responsiveness among low-income groups*. Such a study would provide further evidence both that the data from the U.S. **comports with that found in other countries** and that **the effects of price adjustments would be most pronounced on persons of lower socioeconomic status**.

Choice C, *Kates and Hayward are funded by a sugar-sweetened beverage corporation*, may also have seemed appealing; it suggests that their objections to price adjustments could be motivated by financial considerations rather than actual concern for consumers' autonomy. However, even if Kates and Hayward's motivations were self-serving, that would not disprove the claim that price adjustments threaten consumers' autonomy. Instead, that concern may be legitimate regardless of who raises it. By contrast, **Choice A** would reinforce some of the data cited by the author and would thus support the passage argument.

For an alternative method of explanation based more specifically on passage evidence, you can view this question in the UWorld Qbank (sold separately).

# Passage K: Lengthening the School Day

There may be reasons to reject the idea of lengthening the school day. None of them, however, are *good* reasons. Rather, the supposed demerits of such a proposal fall easily in the face of its numerous financial and social benefits for families.

The greatest of these benefits lies in reducing the need for childcare. It is a curious fact of American life that the adult's work schedule and the child's school schedule are misaligned. Children rise with the sun to head to classes, only to be sent home again hours before parents return from their jobs. In a society where, more often than not, both parents work, this discordance creates the need for an expensive arrangement to fill the gap in families' routines. For instance, studies show that in 2016, childcare costs accounted for 9.5 to 17.5 percent of median family income, depending on the state. Today, 40 percent of families nationwide spend over 15 percent of their income on childcare. Transportation to and from care sites only adds to that expense.

An additional advantage of an extended school day would be to allow for greater diversity and depth in curricula. Schools across the country have increasingly cut instruction in arts, music, and physical education (as well as recess) in order to meet objectives in reading and math. While this unfortunate state of affairs can be partially blamed on overzealous attention to standardized tests, it points to the larger deleterious trend of narrowing students' instruction. With a longer school day, such eliminated subjects can be restored, enriching students with a more well-rounded education.

To this proposal, however, critics may object that the added time would impose strain on educators. Can we truly ask schoolteachers—already among the most overworked individuals in society—to endure even more hours in the classroom? The answer is that a lengthened school day need not distress teachers nor add to their already cumbersome workload. By providing for additional areas of study in the arts and humanities, the extension would give schools cause to hire new, perhaps specialized, faculty to offer these courses. Moreover, the time could also be allocated to sports, academic clubs, and other extracurricular activities.

However, this point speaks to another objection, namely, the cost of adjusting the school day. Whether through paying current teachers more or hiring new ones, implementing such a proposal would entail a significant financial expenditure. There are at least two responses to this line of thinking. First is that this increase in the cost of schooling would be offset and likely surpassed by the aforementioned savings in childcare. Thus, while it is true that schools would require greater funding (likely necessitating higher property taxes), parents would ultimately pay the same or less overall, with greater educational opportunities for their children and fewer transportational burdens. Second is that schools should be better funded regardless. Recently, some schools—especially those in rural areas—have even reduced school weeks to only four days as a cost-saving measure. It is beyond dispute that schools across the board both need and deserve a radically increased investment from citizens. Lengthening the school day is simply one manifestation of how such funding should be utilized.

With this one change, states can coordinate the lives of parents and children, reduce the need for costly childcare, and expand curricular offerings. These worthy and desirable aims provide a clear justification for extending the school day.

*Lengthening the School Day* ©UWorld

Annotations for this passage can be found in the CARS Passage Booklet.

### 3d. External Scenario Support or Challenge Question 3

Which of the following statements, if true, would most *weaken* the passage argument for lengthening the school day?

A. A longer school day's cost in taxes is nearly as much as what parents currently pay for childcare.
B. A longer school day leads to a decreased sense of job satisfaction among teachers.
C. A longer school day increases the amount schools must spend to fund faculty positions.
D. A longer school day produces no improvement in student test scores.

See next page for the *strategy-based explanation* of this question.

> ### 3d. External Scenario Support or Challenge Question 3
> ### Strategy-based Explanation
>
> Which of the following statements, if true, would most *weaken* the passage argument for lengthening the school day?
>
> A. A longer school day's cost in taxes is nearly as much as what parents currently pay for childcare.
> B. A longer school day leads to a decreased sense of job satisfaction among teachers.
> C. A longer school day increases the amount schools must spend to fund faculty positions.
> D. A longer school day produces no improvement in student test scores.

If we think about how the passage is structured, we recall that its argument first offers positive reasons for lengthening the school day, then addresses potential objections to that proposal. Therefore, statements could weaken that argument in one of two ways: by casting doubt on a reason in favor of lengthening the school day, or by casting doubt on the author's response to an objection.

## Applying the Method

*Passage Excerpt*

**[P2] The greatest of these benefits lies in reducing the need for childcare.** It is a curious fact of American life that **the adult's work schedule and the child's school schedule are misaligned.** Children rise with the sun to head to classes, only to be sent home again hours before parents return from their jobs. In a society where, more often than not, both parents work, **this discordance creates the need for an expensive arrangement to fill the gap in families' routines.**

**[P3] An additional advantage of an extended school day would be to allow for greater diversity and depth in curricula.** Schools across the country have increasingly cut instruction in arts, music, and physical education (as well as recess) in order to meet objectives in reading and math. While this unfortunate state of affairs can be partially blamed on overzealous attention to standardized tests, it points to the larger deleterious trend of narrowing students' instruction. **With a longer school day, such eliminated subjects can be restored, enriching students with a more well-rounded education.**

### Step 1: Restate the Relevant Claims or Argument

Aside from its introduction and conclusion, the passage is evenly split: two paragraphs present reasons in favor of lengthening the school day, and two paragraphs respond to objections.

**Points in favor:**

- Lengthening the school day would **reduce the need for childcare**, a problem caused by parents' and children's mismatched schedules.
- Lengthening the school day would **improve the diversity and depth of curricula** because the extra time would allow formerly cut subjects to be restored.

**Responses to objections:**

- Lengthening the school day **need not distress teachers** because schools could hire additional faculty and devote more time to sports, clubs, and other activities.
- Lengthening the school day **would not be too costly** because the increased costs would be offset by childcare savings.

**[P4]** To this proposal, however, critics may object that the added time would impose strain on educators. Can we truly ask schoolteachers—already among the most overworked individuals in society—to endure even more hours in the classroom? The answer is that **a lengthened school day need not distress teachers nor add to their already cumbersome workload.** By providing for additional areas of study in the arts and humanities, the extension would give schools cause to hire new, perhaps specialized, faculty to offer these courses. Moreover, the time could also be allocated to sports, academic clubs, and other extracurricular activities.

**[P5]** However, this point speaks to another objection, namely, the cost of adjusting the school day. Whether through paying current teachers more or hiring new ones, implementing such a proposal would entail a significant financial expenditure. There are at least two responses to this line of thinking. First is that **this increase in the cost of schooling would be offset and likely surpassed by the aforementioned savings in childcare.**

### Step 2: Analyze the Basis of Support

Having restated the passage argument, we can see that the case for lengthening the school day could be weakened in the following ways.

Points in favor would be weakened if:

- lengthening the school day **would not reduce the need for childcare**.
- lengthening the school day **would not create a deeper and more diverse curriculum**.

Responses to objections would be weakened if:

- lengthening the school day **would have a negative impact on teachers**.
- lengthening the school day **would be too costly after all**.

### Step 3: Compare Answer Choices With That Basis

As we turn to the answer choices, one of them clearly reflects what we determined in Step 2: **Choice B**, *A longer school day leads to a decreased sense of job satisfaction among teachers*. If it is true that a longer school day **would have a negative impact on teachers**, then the author's response to that objection would be refuted. Accordingly, the argument for lengthening the school day would be weakened.

Choice A, *A longer school day's cost in taxes is nearly as much as what parents currently pay for childcare*, might also have seemed appealing. However, if the expense of a longer school day were "nearly as much" as what parents currently pay for childcare, that means it would cost slightly less than the childcare it would replace. In that case, the author would be correct in stating that this cost would be **offset by savings in childcare**. Therefore, this statement is consistent with passage claims, whereas **Choice C** would directly contradict part of the passage argument.

For an alternative method of explanation based more specifically on passage evidence, you can view this question in the UWorld Qbank (sold separately).

## Passage I: Meaning: Readers or Authors?

Of late it has become popular among linguists and literary theorists to assert that a work's *meaning* depends upon the individual reader. It is readers, we are told, not authors, who create meaning, by interacting with a text rather than simply receiving it. Thus, a reader transcends the aims of the author, producing their own reading of the text. Indeed, on this line of thinking, even to speak of "the" text is to commit a conceptual error; every text is in fact many texts, a plurality of interpretations that resist comparative evaluation. This view is nonsense. That many otherwise sensible scholars should be attracted to it can perhaps be readily explained, but we should first delineate why the theory goes so far astray....

The absurdity of the view can be demonstrated by a practical analogy. Suppose Smith is conveying his ideas to Jones in conversation (the particular topic is of no consequence). Afterward, we discover that the men differ in their accounts of what Smith had expressed. At this point, Jones may decide that he misunderstood Smith, or perhaps that Smith was unclear. A more complex supposition might be that Smith misused some key term, so his words did not fully match his intentions. Any of these possibilities would reasonably describe why Jones and Smith possessed different opinions about what Smith had said.

What Jones may *not* justifiably conclude is that his own interpretation is what Smith *really meant*. He may not, in effect, say: "Yes, I admit that Smith honestly claims to have been saying something different, but I have formed my own equally correct understanding." Someone who made such an assertion would be suspected of making a joke; if he proved to be serious, we could only conclude that he was deeply confused or else being deliberately quarrelsome. For in questions about what Smith meant, it is surely Smith whose answer must be accepted.... [T]his is not a matter of *agreeing* with a speaker; Jones might judge Smith's ideas to be wrong, unfounded, etc. But whether Smith's ideas are right or wrong is a different matter from what those ideas *are*. On that count, Smith must be the authority.

However, this observation is in no way changed if Smith's ideas are written rather than spoken—sent by letter, for instance. Regardless of any interpretation Jones may produce, the letter's true meaning is whatever Smith intended to convey. Likewise it is, then, with a book, poem, or whatsoever object of literature a scholar (or ordinary reader) encounters. The writing down of ideas does not magically imbue them with malleability or render their content amorphous. From the loftiest tomes of Shakespeare or Milton to the lowliest of yellowed paperbacks, authors produce works with a particular message in mind. It is readers' task to discern that message, not to superimpose their own volitional perspectives.

To think otherwise is to undermine the foundation of literary scholarship. For what is the purpose of such scholarship, if not to seek understanding of an author's creation? One examines the text, taking note of style, historical context, allusions to other works, and other factors, in addition, of course, to the surface sense of the words themselves. If such an enterprise is to be reasonable, it must presume the existence of standards for success: accuracy and inaccuracy, depth or shallowness of analysis, grounds for preferring one interpretation to another. Different readers may come to different conclusions about a text, it is true. But to excise authorial intent from the evaluation of those conclusions does a disservice both to individual works and to literary study as a discipline.

*Meaning: Readers or Authors?* ©UWorld

Annotations for this passage can be found in the CARS Passage Booklet.

### 3d. External Scenario Support or Challenge Question 4

Which of the following statements, if true, would most support the passage author's argument about interpretation?

- A. Novelists often express confusion about the meanings people assign to their works.
- B. Poets often construct verses based on the way words sound regardless of their content.
- C. Psychological studies suggest that authors are often unaware of their own motivations.
- D. Interpretations of famous texts often differ widely between scholars and ordinary readers.

See next page for the *strategy-based explanation* of this question.

Chapter 7: Skill 3 – Reasoning Beyond the Text

> ### 3d. External Scenario Support or Challenge Question 4
> ### Strategy-based Explanation
>
> Which of the following statements, if true, would most support the passage author's argument about interpretation?
>
> A. Novelists often express confusion about the meanings people assign to their works.
> B. Poets often construct verses based on the way words sound regardless of their content.
> C. Psychological studies suggest that authors are often unaware of their own motivations.
> D. Interpretations of famous texts often differ widely between scholars and ordinary readers.

You likely recall that the passage emphasizes a consistent theme about interpretation: that authors rather than readers determine a work's meaning. The argument for this theme develops over several paragraphs, giving us a clear framework from which to analyze it. The correct answer will either reflect the passage author's overall conclusion or reinforce one of the points used to establish it.

## Applying the Method

*Passage Excerpt*

**Step 1: Restate the Relevant Claims or Argument**

[P1] Of late it has become popular among linguists and literary theorists to assert that a work's *meaning* depends upon the individual reader. It is readers, we are told, not authors, who create meaning, by interacting with a text rather than simply receiving it. Thus, a reader transcends the aims of the author, producing their own reading of the text. Indeed, on this line of thinking, even to speak of "the" text is to commit a conceptual error; every text is in fact many texts, a plurality of interpretations that resist comparative evaluation. **This view is nonsense.** That many otherwise sensible scholars should be attracted to it can perhaps be readily explained, but we should first delineate why the theory goes so far astray....

The passage author's argument about interpretation proceeds as follows.

1. It is **nonsense** to think that **readers determine a work's meaning regardless of the author's intention.**

[P3] **What Jones may *not* justifiably conclude is that his own interpretation is what Smith *really meant*.** He may not, in effect, say: "Yes, I admit that Smith honestly claims to have been saying something different, but I have formed my own equally correct understanding." Someone who made such an assertion would be suspected of making a joke; if he proved to be serious, we could only conclude that he was deeply confused

2. A good analogy would be that **in questions about what Smith meant, it is surely Smith whose answer must be accepted**.

316

or else being deliberately quarrelsome. For **in questions about what Smith meant, it is surely Smith whose answer must be accepted**....

[P4] However, **this observation is in no way changed if Smith's ideas are written rather than spoken**—sent by letter, for instance. Regardless of any interpretation Jones may produce, the letter's true meaning is whatever Smith intended to convey. **Likewise it is, then, with a book, poem, or whatsoever object of literature a scholar (or ordinary reader) encounters.** The writing down of ideas does not magically imbue them with malleability or render their content amorphous. From the loftiest tomes of Shakespeare or Milton to the lowliest of yellowed paperbacks, **authors produce works with a particular message in mind. It is readers' task to discern that message**, not to superimpose their own volitional perspectives.

3. This fact **doesn't change when ideas are written rather than spoken**.

4. Therefore, the correct interpretation of a work is **whatever message an author had in mind**, regardless of what individual readers think.

### Step 2: Analyze the Basis of Support

As we have seen, the passage author argues that **readers do not determine the meaning of a written work**, just as listeners do not determine the meaning of what a speaker says.

Accordingly, we can see that this argument would be supported by information that:

- **affirms authors rather than readers** as the source of meaning; or
- **reinforces the analogy** between speakers/authors and listeners/readers.

### Step 3: Compare Answer Choices with That Basis

We determined that the passage author's argument would be supported by information that **affirms authors rather than readers** as the source of meaning or that **reinforces the analogy** that authors are to readers as speakers are to listeners.

Accordingly, the answer choice most consistent with that basis of support is **Choice A**, *Novelists often express confusion about the meanings people assign to their works*. This scenario would suggest that an author hears a reader's interpretation of their work and responds by explaining what they actually meant when writing it. Thus, the implication is that the reader has misinterpreted the work while the author knows its true meaning, just as the passage states: "**authors produce works with a particular message in mind. It is readers' task to discern that message,** not to superimpose their own volitional perspectives."

Choice D might also seem attractive: if *interpretations of famous texts often differ widely between scholars and ordinary readers*, then you might assume that one group is accurately interpreting the text while the other is not. However, this assumption is not justified: the scenario gives no reason to think that *either* group's interpretation is necessarily correct. Therefore, Choice D does not support the author's view of interpretation, leaving **Choice A** as the best answer.

For an alternative method of explanation based more specifically on passage evidence, you can view this question in the UWorld Qbank (sold separately).

# Passage A: The Knights Templar

The seal of the Knights Templar depicts two knights astride a single horse, a visual testament of the order's poverty at its inception in 1119. Nevertheless, these Knights of the Temple—who swore oaths not only of poverty but of chastity, loyalty, and bravery—would eventually become one of the wealthiest and most powerful organizations in the medieval world. So far-reaching was their strength and acclaim that their destruction must have seemed as sudden and surprising as it was utter and irrevocable. The signs of danger could not have been wholly invisible, however, as the Templars' growing influence became perceived as a threat to European rulers.

The first Templars were nine knights who took an oath to defend the Holy Land and any pilgrims who journeyed there after the First Crusade. Having secured a small benefice from Jerusalem's King Baldwin II, the knights inaugurated their mission at the site of the great Temple of Solomon. The order quickly attracted widespread admiration as well as many recruits from crusaders and other knights. Within a year of its founding, the order received a financial endowment from the deeply impressed Count of Anjou, whose example was soon followed by other nobles and monarchs. As early as 1128, the Templars even gained official papal recognition, and their wealth, holdings, and numbers swelled both in the Holy Land and throughout Europe, especially in France and England.

However, this growing power contained the seeds of the order's downfall. Although the Templars were generally held in high esteem, the passage of time saw censure and suspicion directed toward them. The failure of the disastrous siege of Ascalon in 1153 was attributed by some to Templar greed. Similarly, in 1208 Pope Innocent III condemned the wickedness he believed to exist within their ranks. Moreover, their increasingly elevated status brought them into conflict with established authorities. One revealing example occurred in 1252 when, because of the Templars' "many liberties" as well as their "pride and haughtiness," England's King Henry III proposed to curb the order's strength by reclaiming some of its possessions. The Templars' response was unambiguous: "So long as thou dost exercise justice thou wilt reign; but if thou infringe it, thou wilt cease to be King!"

Ultimately, the impoverished Philip the Fair of France joined forces with Pope Clement V to engineer the Templars' downfall beginning in 1307. Conspiring to seize the order's wealth, Clement invited the Templar Master, Jacques de Molay, to meet with him on the pretext of organizing a new crusade to retake the Holy Land. Shortly thereafter, Philip's forces arrested de Molay and his knights on charges the preponderance of which were almost certainly fabricated. Ranging from the mundane to the unspeakably perverse, the accusations even included an incredible entry citing "every crime and abomination that can be committed." Suffering tortures nearly as horrific as the acts of which they were accused, many Templars confessed.

Not everyone believed these charges. Despite Philip's urging, Edward II of England remained convinced that the accusations were false, a view seemingly shared by other rulers. Nevertheless, Clement ordered Edward to extract confessions, a task the king tried to carry out with some measure of mercy. By 1313, the Templar Order had been dissolved by papal decree, and many of its members were dead. The following year, Jacques de Molay was burned at the stake after declaring that the Templars' confessions were lies obtained under torture. Stories would spread that as he died, he condemned Clement and Philip to join him within a year. Although it is impossible to say whether he truly called divine vengeance down upon them, within a few months' time both pope and king had gone to their graves.

*The Knights Templar* ©UWorld

Annotations for this passage can be found in the CARS Passage Booklet.

> **3d. External Scenario Support or Challenge Question 5**
>
> Which of the following statements, if true, would most *challenge* the author's claims about the Knights Templar?
>
> A. The Templars acquired most of their extensive wealth from King Baldwin II.
> B. The Templars had at times given assistance to Philip the Fair before he plotted against them.
> C. The Templars who originally founded their order had previously considered just living as monks in the Holy Land.
> D. The Templars required their recruits to swear that they would never retreat in battle.

See next page for the *strategy-based explanation* of this question.

> ### 3d. External Scenario Support or Challenge Question 5
> ### Strategy-based Explanation
>
> Which of the following statements, if true, would most *challenge* the author's claims about the Knights Templar?
>
> A. The Templars acquired most of their extensive wealth from King Baldwin II.
> B. The Templars had at times given assistance to Philip the Fair before he plotted against them.
> C. The Templars who originally founded their order had previously considered just living as monks in the Holy Land.
> D. The Templars required their recruits to swear that they would never retreat in battle.

This question refers generally to "claims about the Knights Templar" without providing any specifics. Therefore, we cannot follow our usual approach of determining what the answer should look like before turning to the answer choices. In reviewing those choices, we see that they could challenge the author's claims about how the Templars acquired their extensive wealth; their past dealings with Philip the Fair; the original intentions of their founders; or whether they swore never to retreat in battle.

## Applying the Method

*Passage Excerpt*

[P1] The seal of the Knights Templar depicts two knights astride a single horse, a visual testament of **the order's poverty at its inception** in 1119. Nevertheless, these Knights of the Temple—**who swore oaths not only of poverty but of chastity, loyalty, and bravery**—would eventually become one of the wealthiest and most powerful organizations in the medieval world. So far-reaching was their strength and acclaim that their destruction must have seemed as sudden and surprising as it was utter and irrevocable. The signs of danger could not have been wholly invisible, however, as the Templars' growing influence became perceived as a threat to European rulers.

[P2] The first Templars were nine knights who took an oath to defend the Holy Land and any pilgrims who journeyed there after the First Crusade. **Having secured a small benefice from Jerusalem's King Baldwin II**, the knights

**Step 1: Restate the Relevant Claims or Argument**

**Step 2: Analyze the Basis of Support**

Choice A

"The Templars acquired most of their extensive wealth from King Baldwin II."

- The passage describes the Templar order's **poverty at its inception**, when the knights received only **a small benefice from Jerusalem's King Baldwin II**. However, they later received financial support from **the Count of Anjou** as well as **other nobles and monarchs**. As a result, **their wealth, holdings, and numbers swelled**.

Basis of Support

- Based on this information about the source of the Templars' wealth, the author's claims would be *challenged* if: the Templars **were wealthy even when their order was first founded**; or the Templars **acquired their wealth from different sources** than the ones described.

inaugurated their mission at the site of the great Temple of Solomon.  They quickly attracted widespread admiration as well as many recruits from crusaders and other knights.  Within a year of its founding, **the order received a financial endowment from the deeply impressed Count of Anjou, whose example was soon followed by other nobles and monarchs**.  As early as 1128, the Templars even gained official papal recognition, **and their wealth, holdings, and numbers swelled** both in the Holy Land and throughout Europe, especially in France and England.

[P4] Ultimately, the impoverished **Philip the Fair of France joined forces with Pope Clement V to engineer the Templars' downfall** beginning in 1307.  Conspiring to seize the order's wealth, Clement invited the Templar Master, Jacques de Molay, to meet with him on the pretext of organizing a new crusade to retake the Holy Land.  Shortly thereafter, **Philip's forces arrested de Molay and his knights** on charges the preponderance of which were almost certainly fabricated.  Ranging from the mundane to the unspeakably perverse, the accusations even included an incredible entry citing "every crime and abomination that can be committed."  Suffering tortures nearly as horrific as the acts of which they were accused, many Templars confessed.

.

### Choice D

"The Templars required their recruits to swear that they would never retreat in battle."

- The passage states that the Templars **swore oaths of bravery**, which could have included swearing never to retreat in battle.

**Basis of Support**

- This claim would be *challenged* if the Templars **did not really swear such oaths**.

### Choice C:

"The Templars who originally founded their order had previously considered just living as monks in the Holy Land."

- The only claim the passage makes about the Templars' founders is that they **were nine knights who took an oath to defend the Holy Land and any pilgrims who journeyed there after the First Crusade**.

**Basis of Support**

- This claim would be *challenged* if the Templar Order **had been founded in a different way** or if the original founders **had not taken such an oath**.

### Choice B:

"The Templars had at times given assistance to Philip the Fair before he plotted against them."

- According to the passage, **Philip the Fair of France joined forces with Pope Clement V to engineer the Templars' downfall**, and subsequently **arrested Jacques de Molay and his knights**.

**Basis of Support**

- This claim would be *challenged* if Philip **did not conspire to bring down the Templars** or **did not really have them arrested**.

**Step 3: Compare Answer Choices with That Basis**

After considering each answer choice, we can see that only one of them would challenge the author's claims about the Knights Templar: **Choice A**, *The Templars acquired most of their extensive wealth from King Baldwin II*. This statement contradicts the author's claims that the Templars began in **poverty** and received only **a small benefice from Baldwin II**, and also that they later gained wealth from the generosity of **other nobles and monarchs**.

---

### Incompatible or Just Surprising?

As you looked at the answer choices, you might have felt that some of the others didn't fit with passage information. For example, the passage never suggests that the Templars ever *helped Philip the Fair*, nor that their founders *originally considered living as monks*. However, while these statements might be unexpected, the passage doesn't say anything to rule them out. (In fact, both of these statements are historically true—but you don't need to know that to answer the question.)

On the other hand, if Baldwin II gave the Templars only a small amount of financial support, then he *couldn't* have been the source of their extensive wealth. Thus, while it might seem **surprising** if Choices B and C were true, Choice A is actually **incompatible** with passage claims, which is why it would challenge them.

Keeping this surprising/incompatible distinction in mind is helpful for answering many CARS questions that relate to a passage's logical links. Over time, you may find yourself starting to notice these links naturally as you read.

---

For an alternative method of explanation based more specifically on passage evidence, you can view this question in the UWorld Qbank (sold separately).

Chapter 7: Skill 3 – Reasoning Beyond the Text

Lesson 7.5
# RBT Subskill 3e. Applying Passage Perspectives

| Skill 3: Reasoning Beyond the Text ||
|---|---|
| **Subskill** | **Student Objective** |
| 3a. Exemplar Scenario for Passage Claims | Identify which scenario most exemplifies or logically follows from passage claims |
| 3b. Passage Applications to New Context | Apply passage information to a new situation or context |
| 3c. New Claim Support or Challenge | Evaluate how a new claim supports or challenges passage information |
| 3d. External Scenario Support or Challenge | Determine which external scenario supports or challenges passage information |
| **3e. Applying Passage Perspectives** | **Apply the perspective of the author or another source in the passage to a new situation or claim** |
| 3f. Additional Conclusions From New Information | Use new information to draw additional conclusions |
| 3g. Identifying Analogies | Identify analogies or similarities between passage ideas and ideas found outside the passage |

**3e. Applying Passage Perspectives** is the final subskill that focuses on the views of the author or other sources in the passage. For instance, all of the following are **3e** questions:

- The author would probably recommend that political debates should:
- Based on information in the passage, bioethicists would NOT agree with which of the following statements?
- Literary scholarship has always included debate about how the standard interpretation of a work should be determined. Which of the following conceptions is most in accord with the main idea of the passage?

Since these are all examples of Reasoning Beyond the Text questions, their answers come from outside the passage. However, that does not mean that recognizing the correct answer involves guesswork or speculation. Rather, you are extending your knowledge of the passage to find the answer choice that best reflects the source's ideas.

# Method for Answering 3e. Applying Passage Perspectives Questions

## Step 1: Take Note of Viewpoint Indicators

Just as with **1e** and **2e** questions, we can recognize a source's position by paying attention to **Viewpoint Indicators**: words and phrases that reveal perspectives or introduce information that reveals them. Such indicators may function to:

- **express a judgment**

    "Obviously, there is no such thing as a real psychic, and the predictions are nothing but superstition."

- **portray or compare specific views**

    "Yet, where other thinkers equated the ultimate nature of reality with monism, James insisted on a plurality of individual realities."

- **present a conclusion**

    "Thus, what makes a statement a lie is not whether it is false, but whether it is meant to deceive."

By noting these types of words and phrases when they appear, you can quickly identify the beliefs and attitudes held by a particular source in the passage.

## Step 2: Apply the Source's Belief or Attitude

Having recognized the source's belief or attitude, you can then apply that view to the topic the question asks about. While answering the question takes you beyond the passage, the correct answer will still be based on passage information. Note that there is nothing unusual about this process; we frequently think this way in real life.

For example, suppose you were ordering food for a friend who was late to a restaurant. Based on what you know about your friend's likes and dislikes, you could probably choose something they would enjoy even if you had never been to that particular restaurant before. In the same way, the passage tells you what a source believes or feels about a given topic. Thus, based on that knowledge, you can choose the answer that source would agree with, even if its specific contents were never discussed in the passage.

Accordingly, applying a source's view to a new situation or claim is a matter of using what you already know from the passage to determine what else the source would believe or feel.

## Subskill 3e. Applying Passage Perspectives
## (3 Practice Questions)

We now look at three examples from the UWorld Qbank. In each case, the passage and question are first given without commentary, allowing you to practice applying the method yourself. Then, the passage and question are presented again, this time with annotations of the passage (see the CARS Passage Booklet) and a step-by-step explanation of the question using the described method.

# Passage K: Lengthening the School Day

There may be reasons to reject the idea of lengthening the school day. None of them, however, are *good* reasons. Rather, the supposed demerits of such a proposal fall easily in the face of its numerous financial and social benefits for families.

The greatest of these benefits lies in reducing the need for childcare. It is a curious fact of American life that the adult's work schedule and the child's school schedule are misaligned. Children rise with the sun to head to classes, only to be sent home again hours before parents return from their jobs. In a society where, more often than not, both parents work, this discordance creates the need for an expensive arrangement to fill the gap in families' routines. For instance, studies show that in 2016, childcare costs accounted for 9.5 to 17.5 percent of median family income, depending on the state. Today, 40 percent of families nationwide spend over 15 percent of their income on childcare. Transportation to and from care sites only adds to that expense.

An additional advantage of an extended school day would be to allow for greater diversity and depth in curricula. Schools across the country have increasingly cut instruction in arts, music, and physical education (as well as recess) in order to meet objectives in reading and math. While this unfortunate state of affairs can be partially blamed on overzealous attention to standardized tests, it points to the larger deleterious trend of narrowing students' instruction. With a longer school day, such eliminated subjects can be restored, enriching students with a more well-rounded education.

To this proposal, however, critics may object that the added time would impose strain on educators. Can we truly ask schoolteachers—already among the most overworked individuals in society—to endure even more hours in the classroom? The answer is that a lengthened school day need not distress teachers nor add to their already cumbersome workload. By providing for additional areas of study in the arts and humanities, the extension would give schools cause to hire new, perhaps specialized, faculty to offer these courses. Moreover, the time could also be allocated to sports, academic clubs, and other extracurricular activities.

However, this point speaks to another objection, namely, the cost of adjusting the school day. Whether through paying current teachers more or hiring new ones, implementing such a proposal would entail a significant financial expenditure. There are at least two responses to this line of thinking. First is that this increase in the cost of schooling would be offset and likely surpassed by the aforementioned savings in childcare. Thus, while it is true that schools would require greater funding (likely necessitating higher property taxes), parents would ultimately pay the same or less overall, with greater educational opportunities for their children and fewer transportational burdens. Second is that schools should be better funded regardless. Recently, some schools—especially those in rural areas—have even reduced school weeks to only four days as a cost-saving measure. It is beyond dispute that schools across the board both need and deserve a radically increased investment from citizens. Lengthening the school day is simply one manifestation of how such funding should be utilized.

With this one change, states can coordinate the lives of parents and children, reduce the need for costly childcare, and expand curricular offerings. These worthy and desirable aims provide a clear justification for extending the school day.

*Lengthening the School Day* ©UWorld

Annotations for this passage can be found in the CARS Passage Booklet.

### 3e. Applying Passage Perspectives Question 1

The passage author would most likely support a plan for schools to:

   A. enhance classroom instruction through greater focus on STEM disciplines.
   B. spend more classroom time studying particular subjects on certain days.
   C. replace courses in reading and math with courses in the humanities.
   D. require students to take at least one art or music course per semester.

See next page for the *strategy-based explanation* of this question.

## 3e. Applying Passage Perspectives Question 1
## Strategy-based Explanation

The passage author would most likely support a plan for schools to:

    A. enhance classroom instruction through greater focus on STEM disciplines.
    B. spend more classroom time studying particular subjects on certain days.
    C. replace courses in reading and math with courses in the humanities.
    D. require students to take at least one art or music course per semester.

Although this question might seem vague or general at first glance, we have a lot to go on to help answer it. We know the author favors lengthening the school day for the benefits they think would result from such a change. Accordingly, we can conclude that the author would support a plan that would help achieve one or more of those benefits. Thus, the correct answer choice will likely describe such a plan.

## Applying the Method

*Passage Excerpt*

**Step 1: Take Note of Viewpoint Indicators**

[P2] **The greatest of these benefits** lies in reducing the need for childcare. It is a curious fact of American life that the adult's work schedule and the child's school schedule are misaligned. Children rise with the sun to head to classes, only to be sent home again hours before parents return from their jobs. In a society where, more often than not, both parents work, this discordance creates the need for an expensive arrangement to fill the gap in families' routines.

Viewpoint Indicator

According to Paragraph 2, **the greatest benefit** of a lengthened school day would be **reducing the need for childcare**.

[P3] **An additional advantage** of an extended school day would be to allow for greater diversity and depth in curricula. Schools across the country have **increasingly cut instruction in arts, music, and physical education (as well as recess)** in order to meet objectives in reading and math. While **this unfortunate state of affairs** can be partially blamed on **overzealous attention** to standardized tests, it points to the **larger deleterious trend** of narrowing students' instruction. With a longer school day, such eliminated subjects can be restored, **enriching students** with a more well-rounded education.

Viewpoint Indicators

Paragraph 3 describes **an additional advantage** of lengthening the school day: **enriching students** with a more diverse curriculum.

The author specifically mentions **restoring subjects** that have **unfortunately** been eliminated, addressing the **deleterious trend** of removing classes in arts, music, physical education, and recess.

**[P6]** With this one change, states can coordinate the lives of parents and children, reduce the need for costly childcare, and expand curricular offerings. These **worthy and desirable aims** provide a **clear justification** for extending the school day.

**Viewpoint Indicators**

The passage concludes with a summation of the author's view:

Lengthening the school day is **worthy**, **desirable**, and **clearly justified** because it would:

- coordinate the lives of parents and children;
- reduce the need for costly childcare; and
- expand curricular offerings.

### Step 2: Apply the Source's Belief or Attitude

Based on the viewpoint indicators in the passage, we know that the author stresses three major benefits of lengthening the school day: coordinating parents' and children's schedules; reducing the need for childcare; and enabling a more diverse curriculum. Therefore, we can expect the author to **support plans that involve meeting these goals**.

Considering those goals in relation to our question, we can see that only one answer choice would help achieve any of them: **Choice D**, *require students to take at least one art or music course per semester*. This answer reflects the author's view that schools should restore classes in art and music as part of diversifying the curriculum, and is thus the correct choice.

Because the author supports lengthening the school day, Choice B, *spend more classroom time studying particular subjects on certain days*, might also have seemed appealing. However, a closer look at this answer choice shows that it doesn't actually have to do with lengthening the school day. A school could allocate more time to particular subjects on certain days even if the length of the school day remained the same. Accordingly, Choice B doesn't really connect with the author's views, while **Choice D** does.

For an alternative method of explanation based more specifically on passage evidence, you can view this question in the UWorld Qbank (sold separately).

## Passage I: Meaning: Readers or Authors?

Of late it has become popular among linguists and literary theorists to assert that a work's *meaning* depends upon the individual reader. It is readers, we are told, not authors, who create meaning, by interacting with a text rather than simply receiving it. Thus, a reader transcends the aims of the author, producing their own reading of the text. Indeed, on this line of thinking, even to speak of "the" text is to commit a conceptual error; every text is in fact many texts, a plurality of interpretations that resist comparative evaluation. This view is nonsense. That many otherwise sensible scholars should be attracted to it can perhaps be readily explained, but we should first delineate why the theory goes so far astray....

The absurdity of the view can be demonstrated by a practical analogy. Suppose Smith is conveying his ideas to Jones in conversation (the particular topic is of no consequence). Afterward, we discover that the men differ in their accounts of what Smith had expressed. At this point, Jones may decide that he misunderstood Smith, or perhaps that Smith was unclear. A more complex supposition might be that Smith misused some key term, so his words did not fully match his intentions. Any of these possibilities would reasonably describe why Jones and Smith possessed different opinions about what Smith had said.

What Jones may *not* justifiably conclude is that his own interpretation is what Smith *really meant*. He may not, in effect, say: "Yes, I admit that Smith honestly claims to have been saying something different, but I have formed my own equally correct understanding." Someone who made such an assertion would be suspected of making a joke; if he proved to be serious, we could only conclude that he was deeply confused or else being deliberately quarrelsome. For in questions about what Smith meant, it is surely Smith whose answer must be accepted.... [T]his is not a matter of *agreeing* with a speaker; Jones might judge Smith's ideas to be wrong, unfounded, etc. But whether Smith's ideas are right or wrong is a different matter from what those ideas *are*. On that count, Smith must be the authority.

However, this observation is in no way changed if Smith's ideas are written rather than spoken—sent by letter, for instance. Regardless of any interpretation Jones may produce, the letter's true meaning is whatever Smith intended to convey. Likewise it is, then, with a book, poem, or whatsoever object of literature a scholar (or ordinary reader) encounters. The writing down of ideas does not magically imbue them with malleability or render their content amorphous. From the loftiest tomes of Shakespeare or Milton to the lowliest of yellowed paperbacks, authors produce works with a particular message in mind. It is readers' task to discern that message, not to superimpose their own volitional perspectives.

To think otherwise is to undermine the foundation of literary scholarship. For what is the purpose of such scholarship, if not to seek understanding of an author's creation? One examines the text, taking note of style, historical context, allusions to other works, and other factors, in addition, of course, to the surface sense of the words themselves. If such an enterprise is to be reasonable, it must presume the existence of standards for success: accuracy and inaccuracy, depth or shallowness of analysis, grounds for preferring one interpretation to another. Different readers may come to different conclusions about a text, it is true. But to excise authorial intent from the evaluation of those conclusions does a disservice both to individual works and to literary study as a discipline.

*Meaning: Readers or Authors?* ©UWorld

Annotations for this passage can be found in the CARS Passage Booklet.

### 3e. Applying Passage Perspectives Question 2

Based on the passage, a poetry student who encounters an established scholar's interpretation of a poem should do which of the following?

    A. Evaluate the factors that might support or challenge that interpretation.
    B. Develop their own personal interpretation of the poem.
    C. Treat the scholar's greater experience as irrelevant to determining the best interpretation of the poem.
    D. Presume that the scholar's interpretation is probably correct.

See next page for the *strategy-based explanation* of this question.

## 3e. Applying Passage Perspectives Question 2
### Strategy-based Explanation

Based on the passage, a poetry student who encounters an established scholar's interpretation of a poem should do which of the following?

A. Evaluate the factors that might support or challenge that interpretation.
B. Develop their own personal interpretation of the poem.
C. Treat the scholar's greater experience as irrelevant to determining the best interpretation of the poem.
D. Presume that the scholar's interpretation is probably correct.

While the passage does not mention how poetry students should react to scholars' interpretations, it does discuss how we should understand interpretation itself. Therefore, we can consider the question scenario a specific instance of the general ideas presented in the passage. By applying those ideas to the situation described, we can determine what the passage author would say the poetry student should do.

## Applying the Method

*Passage Excerpt*

**[P1]** Of late it has become popular among linguists and literary theorists to assert that a work's *meaning* depends upon the individual reader. It is readers, we are told, not authors, who create meaning, by interacting with a text rather than simply receiving it. Thus, a reader transcends the aims of the author, producing their own reading of the text. Indeed, on this line of thinking, even to speak of "the" text is to commit a conceptual error; every text is in fact many texts, a plurality of interpretations that resist comparative evaluation. **This view is nonsense.** That many otherwise sensible scholars should be attracted to it can perhaps be readily explained, but we should first delineate why the theory goes so far astray....

### Step 1: Take Note of Viewpoint Indicators

**Viewpoint Indicator**

The passage author begins by strongly rejecting a particular view of interpretation.

According to Paragraph 1, it is **nonsense** to think that readers rather than authors create meaning and that interpretations resist evaluation.

Therefore, the passage author believes the opposite: Authors create meaning, and some interpretations are better than others.

**[P5]** To think otherwise is to **undermine the foundation** of literary scholarship. For what is the purpose of such scholarship, if not to seek understanding of an author's creation? One examines the text, taking note of style, historical context, allusions to other works, and other factors, in addition, of course, to the surface sense of the words themselves. **If such an enterprise is to be reasonable**, it must presume the existence of standards for success: accuracy and inaccuracy, depth or shallowness of analysis, grounds for preferring one interpretation to another. Different readers may come to different conclusions about a text, it is true. But to excise authorial intent from the evaluation of those conclusions **does a disservice** both to individual works and to literary study as a discipline.

**Viewpoint Indicators**

The passage author concludes by further distinguishing their own view of interpretation from the one they are arguing against.

According to Paragraph 5, viewing readers as the source of meaning would **undermine the foundation** of literary scholarship. Likewise, viewing all interpretations as equal **does a disservice** both to individual works and to literary study.

Instead, **to be reasonable**, literary study must include grounds for preferring one interpretation to another based on how well each reflects authorial intent.

Therefore, the passage author believes that: interpretations should be judged as better or worse based on how closely they match an author's intended meaning.

### Step 2: Apply the Source's Belief or Attitude

Based on the passage and its viewpoint indicators, we can sum up the passage author's position on interpretation as follows: interpretations can be **more or less accurate** and should be **judged on how successfully they discern an author's intended meaning**.

If we apply that belief to the question scenario, the answer choice that best aligns with it is **Choice A**, *Evaluate the factors that might support or challenge that interpretation*. This answer refers specifically to evaluating the scholar's interpretation, which is consistent with the passage author's emphasis on grounds for preferring one interpretation to another.

Choice C, *Treat the scholar's greater experience as irrelevant to determining the best interpretation of the poem*, may seem plausible as well, since it also refers to evaluating interpretations. However, it is unclear that the passage author would agree with that answer choice as a whole. For example, suppose the scholar had a reputation for accurately identifying that particular poet's authorial intent. In that case, the passage author might think the scholar's past experience was relevant in judging the interpretation. By contrast, the passage author would agree with all of **Choice A**, making it the best answer.

For an alternative method of explanation based more specifically on passage evidence, you can view this question in the UWorld Qbank (sold separately).

## Passage G: Food Costs and Disease

Because frequent consumption of unhealthy foods is strongly linked with cardiometabolic diseases, one way for governments to combat those afflictions may be to modify the eating habits of the general public. Applying economic incentives or disincentives to various types of foods could potentially alter people's diets, leading to more positive health outcomes.

Utilizing national data from 2012 regarding food consumption, health, and economic status, Peñalvo et al. concluded that such price adjustments would help to prevent deaths related to cardiometabolic diseases. According to their analysis, increasing the prices of unhealthy foods such as processed meats and sugary sodas by 10%, while reducing the prices of healthy foods such as fruit and vegetables by 10%, would prevent an estimated 3.4% of yearly deaths in the U.S. Changing prices by 30% would have an even stronger effect, preventing an estimated 9.2% of yearly deaths. This data comports with that found in other countries, such as "previous modeling studies in South Africa and India, where a 20% SSB [sugar-sweetened beverage] tax was estimated to reduce diabetes prevalence by 4% over 20 years." The effects of price adjustments would be most pronounced on persons of lower socioeconomic status, as the researchers "found an overall 18.2% higher price-responsiveness for low versus high SES [socioeconomic status] groups."

This differential effect based on socioeconomic status contributes to concerns about such interventions, however. In *Harvard Public Health Review*, Kates and Hayward ask: "Well intentioned though they may be, at what point do these taxes overstep government influence on an individual's right to autonomy in decision-making? On whom does the increased financial burden of this taxation fall?" They note that taxes on sugar-sweetened beverages, for instance, "are likely to have a greater impact on low-income individuals…because individuals in those settings are more likely to be beholden to cost when making decisions about food."

However, "well intentioned though they may be," the worries that Kates and Hayward express are to some extent misguided. In particular, the idea that taxing unhealthy foods would burden those least able to afford it misses the point. Although the increased taxes would affect anyone who continued to purchase the items despite the higher prices, the goal of raising prices on unhealthy foods is precisely to dissuade people from buying them. As Kates and Hayward themselves remark, "Those in low-income environments may also be the largest consumers of obesogenic foods and therefore most likely to benefit from such a lifestyle change indirectly posed by SSB taxes." As the goal of the taxes is to promote those lifestyle changes, the financial burden objection is a non-starter.

Given this recognition, the question regarding autonomy constitutes a more substantial issue. Nevertheless, that concern also rests on a dubious assumption, as people's autonomy is not necessarily respected in the current situation either. The fact that those of lower socioeconomic status are more likely to have poorer diets suggests that such persons' food choices are the result of financial constraint, not fully autonomous, rational deliberation. Hence, by making healthy foods more affordable relative to unhealthy ones, government intervention might actually facilitate autonomous choices rather than hindering them.

On the other hand, suppose that the disproportionate consumption of unhealthy foods—and associated higher incidence of disease—among certain groups is not the result of financial hardship but rather the result of those persons' perceived self-interest. If so, that would suggest that members of these groups are being encouraged to persist in harmful dietary habits for the sake of corporate profits. In that case, violating autonomy for the sake of health may be permissible, as that practice would be morally preferable to the present system of corporate exploitation.

*Food Costs and Disease* ©UWorld

Annotations for this passage can be found in the CARS Passage Booklet.

Chapter 7: Skill 3 – Reasoning Beyond the Text

## 3e. Applying Passage Perspectives Question 3

Information in the passage suggests that the author would most likely NOT agree with which of the following assertions?

A. For people to make fully autonomous decisions, they need to be free from economic hardship.
B. Preserving people's autonomy is of little concern when considering the good of society.
C. People may be mistaken about what is in their own best interest.
D. When products are taxed, the financial burden typically falls on consumers.

See next page for the *strategy-based explanation* of this question.

Chapter 7: Skill 3 – Reasoning Beyond the Text

## 3e. Applying Passage Perspectives Question 3
## Strategy-based Explanation

Information in the passage suggests that the author would most likely NOT agree with which of the following assertions?

A. For people to make fully autonomous decisions, they need to be free from economic hardship.
B. Preserving people's autonomy is of little concern when considering the good of society.
C. People may be mistaken about what is in their own best interest.
D. When products are taxed, the financial burden typically falls on consumers.

For this question, we are looking for the one answer choice that the author would NOT agree with. Accordingly, by paying attention to viewpoint indicators, we can expect to identify passage claims that are consistent with three of the answer choices, but inconsistent with the correct answer.

## Applying the Method

*Passage Excerpt*

[P4] **As Kates and Hayward themselves remark**, "Those in low-income environments may also be the largest consumers of obesogenic foods and therefore most likely to benefit from such a lifestyle change indirectly posed by SSB taxes." As the goal of the taxes is to promote those lifestyle changes, the financial burden objection is **a non-starter**.

[P5] Given this recognition, the question regarding autonomy constitutes **a more substantial issue**. Nevertheless, that concern also rests on a dubious assumption, as people's autonomy is not necessarily respected in the current situation either. **The fact that** those of lower socioeconomic status are more likely to have poorer diets **suggests that** such persons' food choices are the result of financial constraint, not fully autonomous, rational deliberation. **Hence**, by making healthy foods more affordable relative to unhealthy ones, government intervention might actually *facilitate* autonomous choices rather than hindering them.

### Step 1: Take Note of Viewpoint Indicators

**Viewpoint Indicator**

**Choice D: When products are taxed, the financial burden typically falls on consumers.**

The author agrees that, **as Kates and Hayward themselves remark**, those in low-income environments would be **most impacted by taxes on unhealthy foods**.

Therefore, the author takes for granted that taxing products raises prices for consumers.

**Viewpoint Indicators**

**Choice B: Preserving people's autonomy is of little concern when considering the good of society.**

The passage describes two objections to price adjustments on food. While the author states that the financial burden objection is a **non-starter**, they then assert that the question regarding autonomy constitutes a **more substantial issue**.

In addition, the author later suggests that price adjustments might actually *facilitate* autonomous choices rather than hindering them.

**[P6]** On the other hand, suppose that the disproportionate consumption of unhealthy foods—and associated higher incidence of disease—among certain groups is *not* the result of financial hardship but rather the result of those persons' perceived self-interest. **If so, that would suggest that** members of these groups are being encouraged to persist in harmful dietary habits for the sake of corporate profits. In that case, violating autonomy for the sake of health may be permissible, as that practice would be **morally preferable** to the present system of corporate exploitation.

Therefore, the author appears to treat autonomy as an important consideration.

**Viewpoint Indicator**

**Choice A:** For people to make fully autonomous decisions, they need to be free from economic hardship.

In the view of the author, the evidence **suggests that** some choices are the result of financial constraint, not fully autonomous, rational deliberation.

Thus, the author implies that economic hardship can inhibit a person's autonomy.

**Viewpoint Indicator**

**Choice C:** People may be mistaken about what is in their own best interest.

The author describes a situation **that would suggest that** some people's perceived self-interest is actually harmful to them.

Accordingly, the author believes that people can be mistaken about their own self-interest.

## Step 2: Apply the Source's Belief or Attitude

By paying attention to viewpoint indicators, we saw that the author expresses agreement with three answer choices. They believe that taxing products raises prices for consumers; economic hardship can inhibit a person's autonomy; and people can be mistaken about their own self-interest.

However, the author also states that concern about autonomy is a **more substantial issue** than the financial burden objection, and that adjusting food prices might actually *facilitate* autonomy instead of hindering it. Accordingly, the author appears to treat preserving autonomy as a relevant consideration when arguing for food price adjustments. Consequently, the author would *not* agree with **Choice B**, *Preserving people's autonomy is of little concern when considering the good of society*.

Although the author does state that "violating autonomy for the sake of health may be permissible," their reasoning is that doing so "would be **morally preferable** to the present system of corporate exploitation." In other words, the author believes that violating autonomy is *less wrong* than corporate exploitation; they do not claim that preserving autonomy is of little concern. Therefore, while this statement might initially make us hesitant about **Choice B**, upon reflection we can see that it is the correct answer.

For an alternative method of explanation based more specifically on passage evidence, you can view this question in the UWorld Qbank (sold separately).

Lesson 7.6

# RBT Subskill 3f. Additional Conclusions From New Information

| Skill 3: Reasoning Beyond the Text ||
|---|---|
| **Subskill** | **Student Objective** |
| 3a. Exemplar Scenario for Passage Claims | Identify which scenario most exemplifies or logically follows from passage claims |
| 3b. Passage Applications to New Context | Apply passage information to a new situation or context |
| 3c. New Claim Support or Challenge | Evaluate how a new claim supports or challenges passage information |
| 3d. External Scenario Support or Challenge | Determine which external scenario supports or challenges passage information |
| 3e. Applying Passage Perspectives | Apply the perspective of the author or another source in the passage to a new situation or claim |
| **3f. Additional Conclusions From New Information** | **Use new information to draw additional conclusions** |
| 3g. Identifying Analogies | Identify analogies or similarities between passage ideas and ideas found outside the passage |

**3f. Additional Conclusions From New Information** questions introduce a new fact or assumption, then ask what can be concluded by combining that new information with claims in the passage. Unlike **3c** questions, which also introduce new information (see Lesson 7.3), **3f** questions do not focus on supporting or challenging passage claims. Instead, they may ask:

- what would have happened if a historical event had been different;
- how new data would alter our understanding of a topic;
- the most likely result of a hypothetical action;
- what must have occurred to cause a given phenomenon.

Accordingly, answering these questions is a matter of extrapolating from the combination of passage claims and the new information, to determine what further conclusion can reasonably be drawn.

## Method for Answering 3f. Additional Conclusions From New Information Questions

### Step 1: Consider How the New Information Connects

The first step in answering these questions is to consider how the new information relates to passage claims. There will always be some connection between the two, or else the question would not make sense (see Lesson 7.2 for a similar point about **3b** questions). For example, suppose a passage describes a study in which people who had to solve complicated math problems were then more likely to make errors on an unrelated mental task. Such a passage might include a **3f** question like:

> "A second study found that participants who had just engaged in strenuous exercise were no more likely to make errors on a subsequent mental task. Based on this finding, what would be the most reasonable conclusion?"

Regardless of whatever else the passage discussed, you would see that its connection with the new information was the impact of different activities on mental performance. Accordingly, the answer to the question would reflect that connection.

### Step 2: Incorporate the New Information

Once you have recognized the connection between the new information and passage claims, you are ready to incorporate that new information to determine what conclusion it justifies. In our previous example, you would probably note the following:

> In the first study, prior mental exertion inhibited subsequent mental performance. But in the second study, prior *physical* exertion did not inhibit subsequent mental performance.

Taking these two pieces of information together, a reasonable conclusion would be that *physical fatigue and mental fatigue may be separate phenomena*. Alternatively, a more general conclusion would be that *the relationship between fatigue and performance is more complex than it may appear*. Thus, you could expect the correct answer to express something similar to these conclusions.

### Step 3: Compare the Answer Choices to Your Conclusion

After incorporating the new information into the passage, you will have a clear idea of the kind of statement that would represent the correct answer. One answer choice will reflect this idea, thus constituting an additional conclusion that can be drawn by combining the new and old claims.

## Subskill Connection: 2f vs. 3f

Lesson 6.6 discusses the Reasoning Within the Text subskill **2f. Drawing Additional Inferences**, which concerns using passage information to draw further conclusions. The chief difference between those questions and **3f. Additional Conclusions From New Information** questions is that the latter introduce new information that must be considered.

In a certain sense, however, that difference is artificial. You can think of **3f** questions as asking: "If this information were part of the passage, what could then be concluded?" Accordingly, **3f** questions are not necessarily any more complicated than **2f** questions. Similarly, while Reasoning Beyond the Text questions in general may appear more complicated than others, in some ways this appearance is an illusion.

## Subskill 3f. Additional Conclusions From New Information
### (3 Practice Questions)

We now look at three examples from the UWorld Qbank. In each case, the passage and question are first given without commentary, allowing you to practice applying the method yourself. Then, the passage and question are presented again, this time with annotations of the passage (see the CARS Passage Booklet) and a step-by-step explanation of the question using the described method.

# Passage H: For Whom the Bell Toils

In nineteenth-century America, most people dismissed the notion that someone might assassinate the president. The presumption was based not only on ethics but practicality: a president's term is inherently limited, and an unpopular one could be voted out of office. Therefore, it was reasoned, there would be no need to consider removal through violence. This belief persisted even after the shocking murder of Abraham Lincoln in 1865, which was viewed as an aberration. Thus it was that on July 2, 1881, Charles Guiteau could simply walk up to President James A. Garfield and shoot him in broad daylight. As Richard Menke portrays events, "Guiteau was in fact a madman who had come to identify with a disgruntled wing of the Republican Party after his deranged fantasies of winning a post from the new administration had come to nothing." Believing that God had told him to kill the president, Guiteau thought this act would garner fame for his religious ideas and thereby help to usher in the Apocalypse.

In an interesting parallel, Garfield had felt a sense of divine purpose for his own life after surviving a near-drowning as a young man. Unlike Guiteau's fanatical ravings, however, Garfield's vision worked to the betterment of himself and the world. Candice Millard describes his ascent from extreme poverty to incredible excellence in college, where "by his second year…they made him a professor of literature, mathematics, and ancient languages." Garfield would go on to join the Union Army, where he attained the rank of major general and argued that black soldiers should receive the same pay as their white compatriots. While serving in the war he was nominated for the House of Representatives but accepted the seat only after President Lincoln declared that the country had more need of him as a congressman than as a general. The reluctant politician would later himself become president under similar circumstances, after multiple factions of a deadlocked Republican convention unexpectedly nominated him instead of their original candidates in 1880. An honest man who opposed corruption within the party, Garfield strove both to heal the fractures of the Civil War and to uphold the aims for which it was fought, until "the equal sunlight of liberty shall shine upon every man, black or white, in the Union."

Although Guiteau's bullet would ultimately dim this light for Garfield, the president actually survived the initial attack and for a time appeared headed for recovery. Tragically, however, the hubris shown by his main physician, Dr. Willard Bliss, would lead instead to weeks of prolonged suffering. None of the doctors who examined Garfield were able to locate the bullet, and its lingering presence—along with the unwashed hands of the doctors who probed for it—led to an infection. As the president's condition worsened, inventor Alexander Graham Bell attempted to adapt his patented telephone technology to locate foreign metal in the human body. Inspired by speculation that the bullet's electromagnetic properties might be detectable, Bell used his newly developed "Induction Balance" device to listen for the sounds of electrical interference he hoped would isolate the site of the bullet.

Unfortunately, Bell's searches were unsuccessful. Like Garfield's doctors, he had been looking for the bullet in the wrong area. Menke asserts that Bell's efforts "would probably have fallen short" regardless. However, other historians suggest that Dr. Bliss, unwilling to consider challenges to his original assessment, prevented Bell from more thoroughly searching the president's body. Certainly, Bliss ignored the advice and protestations of other physicians, even as Garfield continued to decline. With death imminent, Garfield asked to be taken to his seaside cottage, where he died on the 19th of September.

*For Whom the Bell Tolls* ©UWorld

Annotations for this passage can be found in the CARS Passage Booklet.

### 3f. Additional Conclusions From New Information Question 1

Which of the following would most likely have occurred if Dr. Bliss had displayed more humility as a physician?

    A. Another doctor would have taken over as Garfield's main physician.
    B. The medical treatment given to Garfield would have been modified.
    C. Alexander Graham Bell would have identified the location of the bullet.
    D. The wound caused by the bullet would not have become infected.

See next page for the *strategy-based explanation* of this question.

> ## 3f. Additional Conclusions From New Information Question 1
> ## Strategy-based Explanation
>
> Which of the following would most likely have occurred if Dr. Bliss had displayed more humility as a physician?
>
>     A. Another doctor would have taken over as Garfield's main physician.
>     B. The medical treatment given to Garfield would have been modified.
>     C. Alexander Graham Bell would have identified the location of the bullet.
>     D. The wound caused by the bullet would not have become infected.

Since the question concerns what would have happened if Dr. Bliss had displayed more humility, the most logical starting point is to ask: what happened because Dr. Bliss did *not* display humility? Based on the answer to that question, we can determine how events might have been different if Dr. Bliss had been more humble.

## Applying the Method

*Passage Excerpt*

[P3] Although Guiteau's bullet would ultimately dim this light for Garfield, the president actually survived the initial attack and for a time appeared headed for recovery. **Tragically, however, the hubris shown by his main physician, Dr. Willard Bliss, would lead instead to weeks of prolonged suffering.** None of the doctors who examined Garfield were able to locate the bullet, and its lingering presence—along with the unwashed hands of the doctors who probed for it—led to an infection. As the president's condition worsened, inventor Alexander Graham Bell attempted to adapt his patented telephone technology to locate foreign metal in the human body. Inspired by speculation that the bullet's electromagnetic properties might be detectable, Bell used his newly developed "Induction Balance" device to listen for the sounds of electrical interference he hoped would isolate the site of the bullet.

[P4] Unfortunately, Bell's searches were unsuccessful. Like Garfield's doctors, he had been looking for the bullet in the wrong area. Menke asserts that Bell's efforts "would probably have fallen short" regardless. However, other historians suggest that Dr. Bliss, unwilling to consider challenges to his

**Step 1: Consider How the New Information Connects**

Dr. Bliss' lack of humility is described in Paragraphs 3 and 4.

First, the passage states that while Garfield initially appeared headed for recovery, the hubris shown by his main physician, Dr. Willard Bliss, would lead instead to weeks of prolonged suffering.

The passage then provides more details, suggesting that Bliss was unwilling to consider challenges to his original assessment of original assessment, prevented Bell from more thoroughly searching the president's body. **Certainly, Bliss ignored the advice and protestations of other physicians, even as Garfield continued to decline.** With death imminent, Garfield asked to be taken to his seaside cottage, where he died on the 19th of September.

> Garfield's condition. Most significantly, he **ignored the advice and protestations of other physicians, even as Garfield continued to decline**.

### Step 2: Incorporate the New Information

The passage conveys that due to Dr. Bliss' lack of humility, he **ignored other doctors' opinions** even though **Garfield continued to suffer instead of recovering**. Therefore, we can reasonably conclude that if Dr. Bliss had instead shown more humility, the opposite would have occurred. In other words, we can draw the following conclusion from the new information: If Dr. Bliss had displayed more humility, **he would have listened to other doctors' opinions** and **Garfield might have recovered.**

### Step 3: Compare the Answer Choices to Your Conclusion

Turning to the answer choices, the one that best matches our conclusion is **Choice B**, *The medical treatment given to Garfield would have been modified*. This answer would reflect Dr. Bliss having **listened to other doctors' opinions** instead of pridefully insisting on his own ineffective ideas about Garfield's treatment.

We might also have considered Choice C, *Alexander Graham Bell would have identified the location of the bullet*. However, while the passage does suggest that Dr. Bliss may have "prevented Bell from more thoroughly searching the president's body," it also suggests that Bell's attempts might have "fallen short" anyway. Thus, although it is *possible* that Bell would have found the bullet, Garfield's treatment would probably have been modified regardless. Therefore, **Choice B** is the better answer.

For an alternative method of explanation based more specifically on passage evidence, you can view this question in the UWorld Qbank (sold separately).

# Passage A: The Knights Templar

The seal of the Knights Templar depicts two knights astride a single horse, a visual testament of the order's poverty at its inception in 1119. Nevertheless, these Knights of the Temple—who swore oaths not only of poverty but of chastity, loyalty, and bravery—would eventually become one of the wealthiest and most powerful organizations in the medieval world. So far-reaching was their strength and acclaim that their destruction must have seemed as sudden and surprising as it was utter and irrevocable. The signs of danger could not have been wholly invisible, however, as the Templars' growing influence became perceived as a threat to European rulers.

The first Templars were nine knights who took an oath to defend the Holy Land and any pilgrims who journeyed there after the First Crusade. Having secured a small benefice from Jerusalem's King Baldwin II, the knights inaugurated their mission at the site of the great Temple of Solomon. The order quickly attracted widespread admiration as well as many recruits from crusaders and other knights. Within a year of its founding, the order received a financial endowment from the deeply impressed Count of Anjou, whose example was soon followed by other nobles and monarchs. As early as 1128, the Templars even gained official papal recognition, and their wealth, holdings, and numbers swelled both in the Holy Land and throughout Europe, especially in France and England.

However, this growing power contained the seeds of the order's downfall. Although the Templars were generally held in high esteem, the passage of time saw censure and suspicion directed toward them. The failure of the disastrous siege of Ascalon in 1153 was attributed by some to Templar greed. Similarly, in 1208 Pope Innocent III condemned the wickedness he believed to exist within their ranks. Moreover, their increasingly elevated status brought them into conflict with established authorities. One revealing example occurred in 1252 when, because of the Templars' "many liberties" as well as their "pride and haughtiness," England's King Henry III proposed to curb the order's strength by reclaiming some of its possessions. The Templars' response was unambiguous: "So long as thou dost exercise justice thou wilt reign; but if thou infringe it, thou wilt cease to be King!"

Ultimately, the impoverished Philip the Fair of France joined forces with Pope Clement V to engineer the Templars' downfall beginning in 1307. Conspiring to seize the order's wealth, Clement invited the Templar Master, Jacques de Molay, to meet with him on the pretext of organizing a new crusade to retake the Holy Land. Shortly thereafter, Philip's forces arrested de Molay and his knights on charges the preponderance of which were almost certainly fabricated. Ranging from the mundane to the unspeakably perverse, the accusations even included an incredible entry citing "every crime and abomination that can be committed." Suffering tortures nearly as horrific as the acts of which they were accused, many Templars confessed.

Not everyone believed these charges. Despite Philip's urging, Edward II of England remained convinced that the accusations were false, a view seemingly shared by other rulers. Nevertheless, Clement ordered Edward to extract confessions, a task the king tried to carry out with some measure of mercy. By 1313, the Templar Order had been dissolved by papal decree, and many of its members were dead. The following year, Jacques de Molay was burned at the stake after declaring that the Templars' confessions were lies obtained under torture. Stories would spread that as he died, he condemned Clement and Philip to join him within a year. Although it is impossible to say whether he truly called divine vengeance down upon them, within a few months' time both pope and king had gone to their graves.

*The Knights Templar* ©UWorld

Annotations for this passage can be found in the CARS Passage Booklet.

## 3f. Additional Conclusions From New Information Question 2

In 1187, Jerusalem fell to an opposing army and the Templar capital was moved to Paris, where it would remain until the order was dissolved. Based on the passage, this information would most reasonably suggest that:

- A. the number of recruits to the Templar Order began to decline.
- B. European nations abandoned their goal of controlling the Holy Land.
- C. fewer knights were willing to embark on crusades.
- D. the Templars would have seemed particularly dangerous to Philip the Fair.

See next page for the *strategy-based explanation* of this question.

Chapter 7: Skill 3 – Reasoning Beyond the Text

> ### 3f. Additional Conclusions From New Information Question 2
> ### Strategy-based Explanation
>
> In 1187, Jerusalem fell to an opposing army and the Templar capital was moved to Paris, where it would remain until the order was dissolved. Based on the passage, this information would most reasonably suggest that:
>
> A. the number of recruits to the Templar Order began to decline.
> B. European nations abandoned their goal of controlling the Holy Land.
> C. fewer knights were willing to embark on crusades.
> D. the Templars would have seemed particularly dangerous to Philip the Fair.

The question provides us with the new information that, for more than a century of its existence, the Templar Order was based in Paris rather than Jerusalem. Accordingly, this information would seem most relevant to passage claims about the Templars' influence in France. By combining those claims with what we now know from the question, we can determine the kind of conclusion that would reasonably follow.

## Applying the Method

*Passage Excerpt*

**[P2]** Within a year of its founding, the order received a financial endowment from the deeply impressed Count of Anjou, whose example was soon followed by other nobles and monarchs. As early as 1128, the Templars even gained official papal recognition, and their wealth, holdings, and numbers **swelled** both in the Holy Land and throughout Europe, **especially in France and England**.

**[P3]** However, **this growing power contained the seeds of the order's downfall**. Although the Templars were generally held in high esteem, the passage of time saw censure and suspicion directed toward them. The failure of the disastrous siege of Ascalon in 1153 was attributed by some to Templar greed. Similarly, in 1208 Pope Innocent III condemned the wickedness he believed to exist within their ranks. Moreover, their increasingly elevated status **brought them into conflict with established authorities**.

**[P4]** Ultimately, the impoverished **Philip the Fair of France** joined forces with Pope Clement V to **engineer the Templars'**

### Step 1: Consider How the New Information Connects

The passage mentions France in connection with the tension that arose between the Templars and other authorities in Europe. For instance, the author tells us that:

1. The Templars' power rapidly **swelled** not only in the Holy Land but also throughout Europe, **especially in France and England**.

2. That same power eventually led to **the order's downfall** as it **brought them into conflict with established authorities**.

3. This **downfall** ultimately occurred when **Philip the Fair of France** joined forces with Pope Clement V.

**downfall** beginning in 1307. Conspiring to seize the order's wealth, Clement invited the Templar Master, Jacques de Molay, to meet with him on the pretext of organizing a new crusade to retake the Holy Land. Shortly thereafter, Philip's forces arrested de Molay and his knights on charges the preponderance of which were almost certainly fabricated.

### Step 2: Incorporate the New Information

We know that the Templars' power led to **conflict with European rulers** and ultimately the **downfall of their order at the hands of a French king**. From the new information introduced by the question, we also know that for most of their existence, the Templars' seat of power was Paris. Therefore, we can reasonably draw the following conclusion: **French rulers would likely have had the most to fear from the Templars' great power**.

### Step 3: Compare the Answer Choices to Your Conclusion

As we look at the answer choices, we can see that one of them is similar to our conclusion from Step 2: **Choice D**, *the Templars would have seemed particularly dangerous to Philip the Fair*. This statement would reasonably follow from what we know about the Templars' power, conflict, and downfall along with the new claim about their capital moving from Jerusalem to Paris.

Choice B, *European nations abandoned their goal of controlling the Holy Land*, may also seem plausible. However, we can rule out this answer for two reasons. First, the fact that the Templars never recaptured their former capital does not mean they didn't *attempt* to do so—they could have simply failed to succeed. Second, even if the Templars had given up on retaking the Holy Land, that does not necessarily mean that European nations also abandoned that goal. Therefore, we cannot reasonably draw this conclusion; by contrast, **Choice D** is well supported.

For an alternative method of explanation based more specifically on passage evidence, you can view this question in the UWorld Qbank (sold separately).

## Passage L: When Defense Is Indefensible

Suppose a prosecutor is considering whether to bring a case to trial. He is not sure that the suspect is guilty—in fact, based on the evidence, it's more likely that the suspect is *not* guilty. Nevertheless, he feels confident he can secure a guilty verdict. His powers of persuasion are considerable, and there's a good chance he could trick a jury into believing the evidence is strong instead of weak. In addition, the case is high profile and could be very lucrative; winning would likely lead to a substantial raise or promotion. He decides to charge the suspect, and ultimately succeeds in persuading the jury to convict.

Looking at this situation, most of us would easily judge the prosecutor as extremely unethical. His conduct is outrageous and wrong—he clearly acted with corrupt intent, perpetrating injustice in order to profit financially. Why is it not shocking, then, that we tolerate the mirror image of this behavior from defense attorneys? For they engage in the same outrageous conduct, only on the other side. Paid handsomely to represent even the vilest of clients, they apply their oratorical prowess to manipulating jury perception, keeping the guilty free and unpunished in exchange for money and status. To the extent that this behavior takes place, are some defense attorneys as unethical as our hypothetical prosecutor?

It is worth distinguishing two senses of the word "ethical" here. For there are standards of *professional* ethics to which any attorney must conform, including standards particular to the defense. Most relevant to our purposes, a lawyer is obligated to provide their client with a "zealous defense." In other words, once an attorney takes on a client, they are ethically bound to promote that client's rights, interests, or innocence—in fact, *not* to do so would be *unethical*. Thus, one might try to suggest that this obligation undermines the claim that some defense attorneys act unethically.

However, meeting that professional standard is not the same as being ethical in the general sense of the word. The standard depends on the condition: *once an attorney takes on a client*. With the exception of court-appointed attorneys or public defenders, who are assigned to provide representation to those who would otherwise lack it, an attorney is never required to represent a defendant. Therefore, meeting one's obligations as a defense attorney does not necessarily make one ethical, because the choice to accept a specific case (and thus to incur those obligations in the first place) may itself be an unethical act.

Moreover, the role of court-appointed attorneys is to help protect the rights of citizens who cannot secure their own representation, usually for financial reasons. Although preserving those rights is necessary to uphold justice, this situation highlights how wealth and class contribute to *injustice*. While some defendants possess the means to hire top-level private lawyers, others must depend on public servants—frequently less experienced lawyers from overloaded, understaffed agencies. As a result, the rich are more likely to escape conviction even when they are guilty, and the poor are more likely to be convicted even when they are innocent.

It is doubtful that private defense attorneys could be somehow forbidden from representing guilty clients. Hence, the needed reforms to the system must come from individual attorneys committing to work for the right reasons. For those who strive to ensure citizens' rights, or who truly believe their clients are innocent, providing a defense is a noble undertaking. But for those whose overriding motivation is greed, that legally "zealous defense" is ethically indefensible.

*When Defense Is Indefensible* ©UWorld

Annotations for this passage can be found in the CARS Passage Booklet.

### 3f. Additional Conclusions From New Information Question 3

If statistics suggest that the effectiveness of a legal defense correlates positively with the expense of securing it, which of the following conclusions would be reasonable?

   I. The most talented defense attorneys tend to be hired by private firms rather than government agencies.
   II. Increasing the funding available to public defenders would give disadvantaged citizens more effective representation.
   III. Highly paid lawyers face more professional obligations in taking on clients than lesser paid lawyers.

A. I only
B. III only
C. I and II only
D. II and III only

See next page for the *strategy-based explanation* of this question.

## 3f. Additional Conclusions From New Information Question 3 Strategy-based Explanation

If statistics suggest that the effectiveness of a legal defense correlates positively with the expense of securing it, which of the following conclusions would be reasonable?

   I. The most talented defense attorneys tend to be hired by private firms rather than government agencies.
   II. Increasing the funding available to public defenders would give disadvantaged citizens more effective representation.
   III. Highly paid lawyers face more professional obligations in taking on clients than lesser paid lawyers.

A. I only
B. III only
C. I and II only
D. II and III only

The question proposes a correlation between the effectiveness of a legal defense and how expensive it is. Thus, by applying this relationship to passage information on the quality and cost of legal defenses, we can determine the most reasonable conclusions about: *the most talented defense attorneys* (Option I); *disadvantaged citizens' representation* (Option II); and *lawyers' professional obligations* (Option III).

## Applying the Method

*Passage Excerpt*

[P3] However, meeting that professional standard is not the same as being ethical in the general sense of the word. The standard depends on the condition: *once an attorney takes on a client*. **With the exception of court-appointed attorneys or public defenders**, who are assigned to provide representation to those who would otherwise lack it, **an attorney is never required to represent a defendant.** Therefore, meeting one's obligations as a defense attorney does not necessarily make one ethical, because the choice to accept a specific case (and thus to incur those obligations in the first place) may itself be an unethical act.

[P4] Moreover, **the role of court-appointed attorneys is to help protect the rights of citizens who cannot secure their own representation, usually for financial reasons**. Although

### Step 1: Consider How the New Information Connects

The passage discusses the quality and cost of legal representation by comparing private and public attorneys.

**Private attorneys**:
- are often **top-level lawyers**.
- serve only those with **the means to hire them**.
- are **never required to represent a defendant**.

preserving those rights is necessary to uphold justice, this situation highlights how wealth and class contribute to *injustice*. While some defendants possess **the means to hire top-level private lawyers**, others must depend on **public servants— frequently less experienced lawyers from overloaded, understaffed agencies**. As a result, **the rich are more likely to escape conviction** even when they are guilty, and **the poor are more likely to be convicted** even when they are innocent.

**Public attorneys:**

- are often **less experienced lawyers from overloaded, understaffed agencies**.
- serve those who cannot secure their own representation, **usually for financial reasons**.
- are **required to defend anyone** to whom they are assigned.

### Step 2: Incorporate the New Information

If there is a **positive correlation between the effectiveness and the expense of legal defenses**, then we can draw the following conclusions from passage information.

As we have seen, the passage draws a distinction between expensive, **top-level private lawyers** and the public servants who are **frequently less experienced lawyers from overloaded, understaffed agencies**. Therefore, the correlation described **would reinforce this distinction** between private and public lawyers, supporting the claim that *the most talented defense attorneys tend to be hired by private firms rather than government agencies*. **(Option I)**

In addition, the fact that public attorneys tend to work for **overloaded, understaffed agencies** implies that those agencies have insufficient resources. Therefore, we can reasonably conclude that providing more resources to such agencies would help them better represent their clients. Thus, the correlation described **would support the claim that** *increasing the funding available to public defenders would give disadvantaged citizens more effective representation.* **(Option II)**

Finally, the highly paid private attorneys are **never required to represent a defendant**, while the lesser paid public attorneys are **required to defend anyone** to whom they are assigned. Accordingly, lesser paid attorneys face professional obligations that higher paid attorneys do not face. Since this difference in obligations is not based on how expensive or effective a defense is, the correlation described **is irrelevant to that difference**. Consequently, we can *rule out* the claim that *highly paid lawyers face more professional obligations in taking on clients than lesser paid lawyers.* (Option III)

### Step 3: Compare the Answer Choices to Your Conclusion

By incorporating **the correlation described in the question** with passage information, we concluded that both **Option I** and **Option II** should be part of the correct answer, but Option III should not. Thus, the correct answer should be *I and II only*. Looking at the answer choices, we see that this combination does appear. Therefore, we can feel confident in our reasoning: the correct answer is **Choice C**, *I and II only*.

For an alternative method of explanation based more specifically on passage evidence, you can view this question in the UWorld Qbank (sold separately).

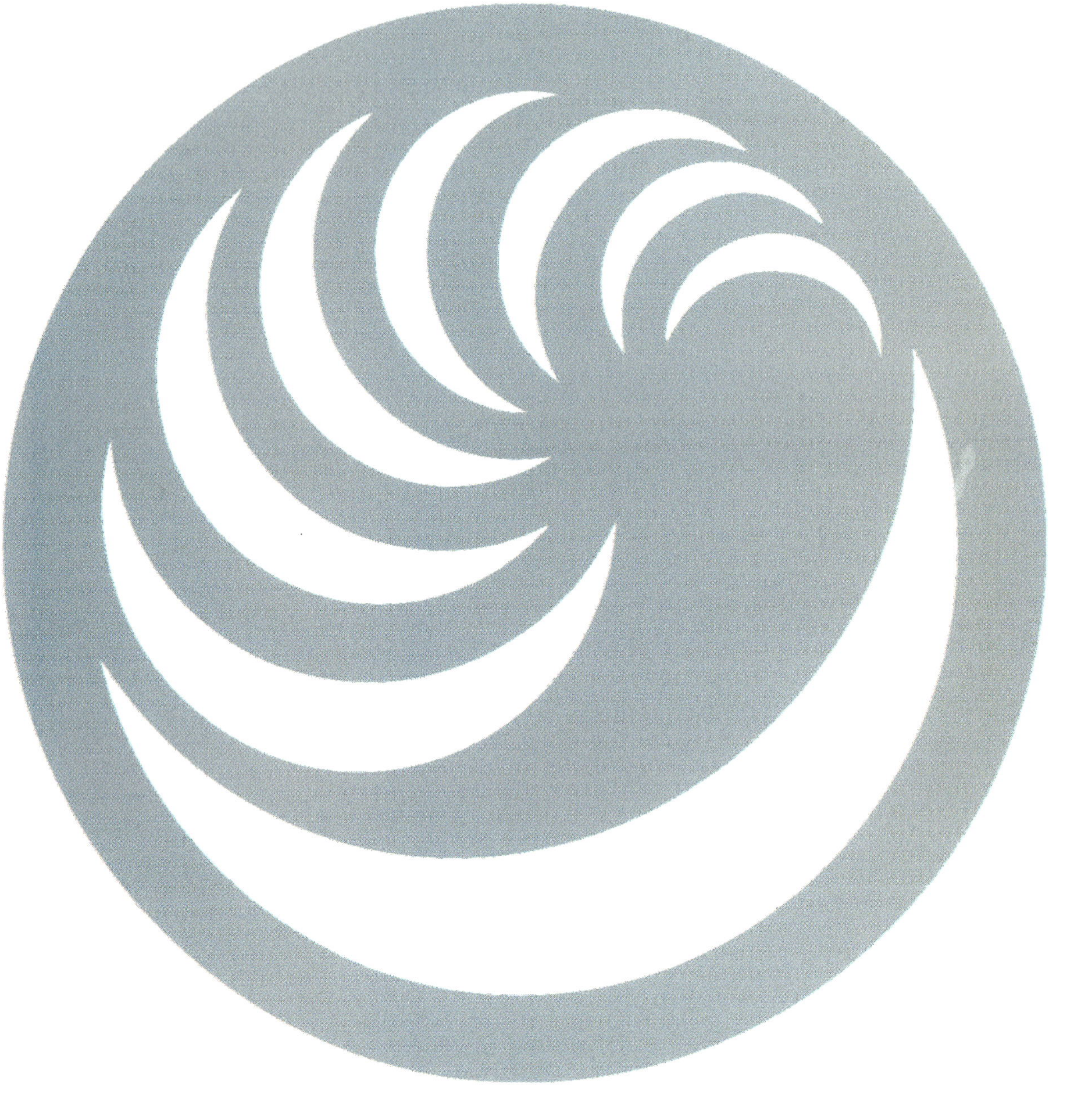

Lesson 7.7

# RBT Subskill 3g. Identifying Analogies

| Skill 3: Reasoning Beyond the Text ||
|---|---|
| **Subskill** | **Student Objective** |
| 3a. Exemplar Scenario for Passage Claims | Identify which scenario most exemplifies or logically follows from passage claims |
| 3b. Passage Applications to New Context | Apply passage information to a new situation or context |
| 3c. New Claim Support or Challenge | Evaluate how a new claim supports or challenges passage information |
| 3d. External Scenario Support or Challenge | Determine which external scenario supports or challenges passage information |
| 3e. Applying Passage Perspectives | Apply the perspective of the author or another source in the passage to a new situation or claim |
| 3f. Additional Conclusions From New Information | Use new information to draw additional conclusions |
| **3g. Identifying Analogies** | **Identify analogies or similarities between passage ideas and ideas found outside the passage** |

The final subskill for Reasoning Beyond the Text questions is **3g. Identifying Analogies**. Although analogy questions may sometimes seem to be overly abstract or to require an unfamiliar way of thinking, they can actually be some of the most straightforward types of question you will encounter in CARS. By approaching these questions in an organized way, answering them becomes a simple process.

# Method for Answering 3g. Identifying Analogies Questions

## Step 1: Identify the Relationship Between Elements

Fundamentally, an analogy is just a comparison between two entities or concepts. In CARS, analogy questions typically refer to an idea in the passage, then ask which external idea it is most like. The key to answering these questions is to at first ignore the comparison being made, and instead just **analyze the original idea: if you break it down into its individual elements, how are those elements related**?

For example, suppose a passage describes a loan repayment plan that begins at a low interest rate but then greatly increases over time. If we think about this situation, we can note that it:

- starts off easy, but later grows difficult;
- is appealing at first, but dangerous in the long run.

Thus, these descriptions express how the earlier and later parts of the loan repayment plan relate to each other.

## Step 2: Match Answer Choices to That Relationship

Once you have identified the relationship essential to the original idea, the next step is to **apply that relationship to the answer choices**. Continuing with our previous example, suppose you encountered a question like: "Which of the following situations for a kayaker would be most similar to the loan repayment plan for a borrower?"

The correct answer might be: *a slow moving river that eventually leads to rapids and waterfalls*. Although kayaking down a river and repaying a loan have seemingly little in common, the basic elements of the two situations are still the same: like the loan repayment plan, the river starts off easy and appealing, then grows difficult and dangerous over time.

Accordingly, even when the passage scenario and the outside scenario seem very different, you can recognize the correct answer by matching its elements with those of the previously identified relationship.

## Subskill 3g. Identifying Analogies
### (3 Practice Questions)

We now look at three examples from the UWorld Qbank. In each case, the passage and question are first given without commentary, allowing you to practice applying the method yourself. Then, the passage and question are presented again, this time with annotations of the passage (see the CARS Passage Booklet) and a step-by-step explanation of the question using the described method.

# Passage L: When Defense Is Indefensible

Suppose a prosecutor is considering whether to bring a case to trial. He is not sure that the suspect is guilty—in fact, based on the evidence, it's more likely that the suspect is *not* guilty. Nevertheless, he feels confident he can secure a guilty verdict. His powers of persuasion are considerable, and there's a good chance he could trick a jury into believing the evidence is strong instead of weak. In addition, the case is high profile and could be very lucrative; winning would likely lead to a substantial raise or promotion. He decides to charge the suspect, and ultimately succeeds in persuading the jury to convict.

Looking at this situation, most of us would easily judge the prosecutor as extremely unethical. His conduct is outrageous and wrong—he clearly acted with corrupt intent, perpetrating injustice in order to profit financially. Why is it not shocking, then, that we tolerate the mirror image of this behavior from defense attorneys? For they engage in the same outrageous conduct, only on the other side. Paid handsomely to represent even the vilest of clients, they apply their oratorical prowess to manipulating jury perception, keeping the guilty free and unpunished in exchange for money and status. To the extent that this behavior takes place, are some defense attorneys as unethical as our hypothetical prosecutor?

It is worth distinguishing two senses of the word "ethical" here. For there are standards of *professional* ethics to which any attorney must conform, including standards particular to the defense. Most relevant to our purposes, a lawyer is obligated to provide their client with a "zealous defense." In other words, once an attorney takes on a client, they are ethically bound to promote that client's rights, interests, or innocence—in fact, *not* to do so would be *unethical*. Thus, one might try to suggest that this obligation undermines the claim that some defense attorneys act unethically.

However, meeting that professional standard is not the same as being ethical in the general sense of the word. The standard depends on the condition: *once an attorney takes on a client*. With the exception of court-appointed attorneys or public defenders, who are assigned to provide representation to those who would otherwise lack it, an attorney is never required to represent a defendant. Therefore, meeting one's obligations as a defense attorney does not necessarily make one ethical, because the choice to accept a specific case (and thus to incur those obligations in the first place) may itself be an unethical act.

Moreover, the role of court-appointed attorneys is to help protect the rights of citizens who cannot secure their own representation, usually for financial reasons. Although preserving those rights is necessary to uphold justice, this situation highlights how wealth and class contribute to *injustice*. While some defendants possess the means to hire top-level private lawyers, others must depend on public servants—frequently less experienced lawyers from overloaded, understaffed agencies. As a result, the rich are more likely to escape conviction even when they are guilty, and the poor are more likely to be convicted even when they are innocent.

It is doubtful that private defense attorneys could be somehow forbidden from representing guilty clients. Hence, the needed reforms to the system must come from individual attorneys committing to work for the right reasons. For those who strive to ensure citizens' rights, or who truly believe their clients are innocent, providing a defense is a noble undertaking. But for those whose overriding motivation is greed, that legally "zealous defense" is ethically indefensible.

*When Defense Is Indefensible* ©UWorld

Annotations for this passage can be found in the CARS Passage Booklet.

### 3g. Identifying Analogies Question 1

Based on the author's description, the behavior of the prosecutor in Paragraph 1 is most like that of:

A. a college admissions officer who ignores a student's personal essay so she can focus on judging academic performance.
B. a hiring manager who sabotages a more qualified candidate's application so she can give the job to her friend.
C. a casting director who sabotages a more talented actor's audition so she can take the part herself.
D. a restaurant manager who fires her head chef so she can hire a more experienced one to improve her menu.

See next page for the *strategy-based explanation* of this question.

> ### 3g. Identifying Analogies Question 1
> ### Strategy-based Explanation
>
> Based on the author's description, the behavior of the prosecutor in Paragraph 1 is most like that of:
>
> A. a college admissions officer who ignores a student's personal essay so she can focus on judging academic performance.
> B. a hiring manager who sabotages a more qualified candidate's application so she can give the job to her friend.
> C. a casting director who sabotages a more talented actor's audition so she can take the part herself.
> D. a restaurant manager who fires her head chef so she can hire a more experienced one to improve her menu.

The question asks us to determine which hypothetical person acts most like the prosecutor described in the passage. Accordingly, we must analyze the prosecutor's behavior to identify the general elements of his situation. Then, we can recognize the correct answer as the scenario that includes those same elements.

## Applying the Method

*Passage Excerpt*

**[P1]** Suppose a prosecutor is considering whether to bring a case to trial. He is not sure that the suspect is guilty—in fact, based on the evidence, **it's more likely that the suspect is not guilty.** Nevertheless, he feels confident he can secure a guilty verdict. His powers of persuasion are considerable, and **there's a good chance he could trick a jury** into believing the evidence is strong instead of weak. In addition, the case is high profile and **could be very lucrative**; winning would likely lead to a substantial raise or promotion. **He decides to charge the suspect**, and ultimately succeeds in **persuading the jury to convict.**

### Step 1: Identify the Relationship Between Elements

Reviewing Paragraph 1, we see that the prosecutor's scenario includes the following features:

- The prosecutor knows the suspect is probably not guilty;
- However, he would personally gain if he gets a conviction;
- So, he tricks the jury into convicting the suspect anyway.

In a more general sense, the structure of that scenario is: **A unjustly harms B so that A will personally gain.**

## Step 2: Match Answer Choices to That Relationship

As we have seen, the hypothetical prosecutor's situation can be described as: **A unjustly harms B so that A will personally gain**. Accordingly, the correct answer will represent a relationship with this same structure.

Turning to the answer choices, we find this relationship exemplified in **Choice C**, *a casting director who sabotages a more talented actor's audition so she can take the part herself*. If the actor is more talented then they are more deserving of the part; thus, the casting director treats the actor unjustly. Moreover, the casting director gains personally by taking the part for herself. Consequently, her behavior parallels that of the prosecutor, making **Choice C** correct.

We might have also considered Choice B, *a hiring manager who sabotages a more qualified candidate's application so she can give the job to her friend*. However, while this scenario does involve treating someone unjustly, it is not the hiring manager but her friend who stands to gain from that unjust behavior. Thus, we now have a third party: A unjustly harms B so that C will gain. Therefore, this answer choice reflects only some elements of the prosecutor's situation, leaving **Choice C** as the better answer.

For an alternative method of explanation based more specifically on passage evidence, you can view this question in the UWorld Qbank (sold separately).

## Passage I: Meaning: Readers or Authors?

Of late it has become popular among linguists and literary theorists to assert that a work's *meaning* depends upon the individual reader. It is readers, we are told, not authors, who create meaning, by interacting with a text rather than simply receiving it. Thus, a reader transcends the aims of the author, producing their own reading of the text. Indeed, on this line of thinking, even to speak of "the" text is to commit a conceptual error; every text is in fact many texts, a plurality of interpretations that resist comparative evaluation. This view is nonsense. That many otherwise sensible scholars should be attracted to it can perhaps be readily explained, but we should first delineate why the theory goes so far astray....

The absurdity of the view can be demonstrated by a practical analogy. Suppose Smith is conveying his ideas to Jones in conversation (the particular topic is of no consequence). Afterward, we discover that the men differ in their accounts of what Smith had expressed. At this point, Jones may decide that he misunderstood Smith, or perhaps that Smith was unclear. A more complex supposition might be that Smith misused some key term, so his words did not fully match his intentions. Any of these possibilities would reasonably describe why Jones and Smith possessed different opinions about what Smith had said.

What Jones may *not* justifiably conclude is that his own interpretation is what Smith *really meant*. He may not, in effect, say: "Yes, I admit that Smith honestly claims to have been saying something different, but I have formed my own equally correct understanding." Someone who made such an assertion would be suspected of making a joke; if he proved to be serious, we could only conclude that he was deeply confused or else being deliberately quarrelsome. For in questions about what Smith meant, it is surely Smith whose answer must be accepted…. [T]his is not a matter of *agreeing* with a speaker; Jones might judge Smith's ideas to be wrong, unfounded, etc. But whether Smith's ideas are right or wrong is a different matter from what those ideas *are*. On that count, Smith must be the authority.

However, this observation is in no way changed if Smith's ideas are written rather than spoken—sent by letter, for instance. Regardless of any interpretation Jones may produce, the letter's true meaning is whatever Smith intended to convey. Likewise it is, then, with a book, poem, or whatsoever object of literature a scholar (or ordinary reader) encounters. The writing down of ideas does not magically imbue them with malleability or render their content amorphous. From the loftiest tomes of Shakespeare or Milton to the lowliest of yellowed paperbacks, authors produce works with a particular message in mind. It is readers' task to discern that message, not to superimpose their own volitional perspectives.

To think otherwise is to undermine the foundation of literary scholarship. For what is the purpose of such scholarship, if not to seek understanding of an author's creation? One examines the text, taking note of style, historical context, allusions to other works, and other factors, in addition, of course, to the surface sense of the words themselves. If such an enterprise is to be reasonable, it must presume the existence of standards for success: accuracy and inaccuracy, depth or shallowness of analysis, grounds for preferring one interpretation to another. Different readers may come to different conclusions about a text, it is true. But to excise authorial intent from the evaluation of those conclusions does a disservice both to individual works and to literary study as a discipline.

*Meaning: Readers or Authors?* ©UWorld

Annotations for this passage can be found in the CARS Passage Booklet.

### 3g. Identifying Analogies Question 2

Based on the view of interpretation described by the passage author in Paragraph 1, which of the following areas of study would be most similar to literary scholarship?

- A. Code decryption
- B. Linguistic translation
- C. Art criticism
- D. Forensic accounting

See next page for the *strategy-based explanation* of this question.

> ### 3g. Identifying Analogies Question 2
> ### Strategy-based Explanation
>
> Based on the view of interpretation described by the passage author in Paragraph 1, which of the following areas of study would be most similar to literary scholarship?
>
> A. Code decryption
> B. Linguistic translation
> C. Art criticism
> D. Forensic accounting

We could rephrase the question as: "The view of interpretation described in Paragraph 1 could most reasonably be applied to which area of study?" Accordingly, our task is to review that description, then consider its potential fit with each answer choice. Most likely, the correct answer will represent the only listed discipline for which such a view of interpretation might make sense.

## Applying the Method

*Passage Excerpt*

[P1] Of late it has become popular among linguists and literary theorists to assert that **a work's *meaning* depends upon the individual reader. It is readers**, we are told, **not authors, who create meaning**, by interacting with a text rather than simply receiving it. Thus, **a reader transcends the aims of the author**, producing their own reading of the text. Indeed, on this line of thinking, even to speak of "the" text is to commit a conceptual error; **every text is in fact many texts, a plurality of interpretations that resist comparative evaluation**. This view is nonsense. That many otherwise sensible scholars should be attracted to it can perhaps be readily explained, but we should first delineate why the theory goes so far astray….

### Step 1: Identify the Relationship Between Elements

Paragraph 1 describes a view of interpretation within the field of literary scholarship in which meaning is determined by readers and the author's view is irrelevant. On this way of thinking, multiple readers create various interpretations that cannot be evaluated as better or worse and are thus each correct for the person who creates them.

Therefore, according to this view of literary scholarship, **the meaning of a literary work is whatever each reader interprets it to mean**.

In more general terms, this view of interpretation proposes the following relationship: **a work's meaning is whatever any individual believes it to be**.

### Step 2: Match Answer Choices to That Relationship

On the view of interpretation described in Paragraph 1, **a work's meaning is whatever any individual believes it to be**. Turning to the answer choices, we find another area in which interpretations of works might vary: **Choice C**, *art criticism*. Just as literary scholarship analyzes the meaning of literary works, art criticism analyzes the meaning of artworks. Furthermore, different observers could clearly interpret the same artwork in different ways. Accordingly, the proposed relationship between a work and its meaning could easily be applied to art in the same way it has been applied to literature, making **Choice C** the correct answer.

You probably noticed that the author actually *disagrees* with this view of interpretation, calling it "nonsense" in Paragraph 1. However, the question does not ask about the author's own view of interpretation but rather about the view they describe in the passage. Thus, if we do not distinguish between what the author believes and a view they merely discuss, we would likely rule out the right answer by mistake. By being careful about both what the question asks and what the passage says, however, we can recognize that **Choice C** is correct.


For an alternative method of explanation based more specifically on passage evidence, you can view this question in the UWorld Qbank (sold separately).

## Passage D: American Local Motives

Locomotives were invented in England, with the first major railroad connecting Liverpool and Manchester in 1830. However, it was in America that railroads would be put to the greatest use in the nineteenth century. On May 10, 1869, the Union Pacific and Central Pacific lines met at Promontory Point, Utah, joining from opposite directions to complete a years-long project—the Transcontinental Railroad. This momentous event connected the eastern half of the United States with its western frontier and facilitated the construction of additional lines in between. As a result, journeys that had previously taken several months by horse and carriage now required less than a week's travel. By 1887 there were nearly 164,000 miles of railroad tracks in America, and by 1916 that number had swelled to over 254,000.

While the United States still has the largest railroad network in the world, it operates largely in the background of American life, and citizens no longer view trains with the sense of importance those machines once commanded. Nevertheless, the economic and industrial advantages those citizens enjoy today would not have been possible without America's history of trains; as Tom Zoellner reminds us, "Under the skin of modernity lies a skeleton of railroad tracks." Although airplanes and automobiles have now assumed greater prominence, the time has arrived for the resurgence of railroads. A revitalized and advanced railway system would confer numerous essential benefits on both the United States and the globe.

The chief obstacles to garnering support for such a project are the current dominance of the automobile and the languishing technology of existing railroads. In a sense these two obstacles are one, as American dependence on personal automobiles is partially due to the paucity of rapid public transportation. The railroads of Europe and Japan, by comparison, have vastly outpaced their American counterparts. Japan has operated high-speed rail lines continuously since 1964, and in 2007, a French train set a record of 357 miles per hour. While that speed was achieved under tightly controlled conditions, it still speaks to the great disparity in railroad development between the United States and other countries since the mid-twentieth century. British trains travel at speeds much higher than those in America, where both the trains themselves and the infrastructure to support them have simply been allowed to fall behind. In much of Europe it is common for trains to travel at close to 200 miles per hour.

To invest in a modern network of railroads would improve the United States in much the same way that the first railroads did in the nineteenth and early twentieth centuries. A high-speed passenger rail system would dramatically transform American life as travel between cities and states became quicker and more convenient, encouraging commerce, business, and tourism. Such a system would also make important strides in environmental preservation. According to a 2007 British study, "CO2 emissions from aircraft operations are...at least five times greater" than those from high-speed trains. For similar reasons, Osaka, Japan, was ranked as "the best...green transportation city in Asia" by the 2011 *Green City Index*. As Lee-in Chen Chiu notes in *The Kyoto Economic Review*, Osakans travel by railway more than twice as much as they travel by car.

It is true that developing a countrywide high-speed rail system would come with significant costs. However, that was also true of the original Transcontinental Railroad, as indeed it is with virtually any great project undertaken for the public good. We should thus move ahead with confidence that the rewards will outweigh the expenditure as citizens increasingly choose to travel by train. Both for society's gain and the crucial well-being of the planet, our path forward should proceed upon rails.

*American Local Motives* ©UWorld

Annotations for this passage can be found in the CARS Passage Booklet.

## 3g. Identifying Analogies Question 3

Which of the following is most like the relationship between American railroads and British railroads as they are described in the passage?

A. An apprentice artist surpasses a master's early works but does not live up to the master's later works.
B. A financial investor earns large sums of money in the short term but loses a larger amount over time.
C. A rookie athlete sets a team record but in later seasons his record is broken by a newer player.
D. An investigative journalist races to break a story but is beaten to the punch by a rival network.

See next page for the *strategy-based explanation* of this question.

## 3g. Identifying Analogies Question 3
## Strategy-based Explanation

Which of the following is most like the relationship between American railroads and British railroads as they are described in the passage?

A. An apprentice artist surpasses a master's early works but does not live up to the master's later works.
B. A financial investor earns large sums of money in the short term but loses a larger amount over time.
C. A rookie athlete sets a team record but in later seasons his record is broken by a newer player.
D. An investigative journalist races to break a story but is beaten to the punch by a rival network.

Unlike the previous two questions, this one does not refer to a specific section of the passage but merely alludes to how American and British railroads are described. Therefore, the relevant information may come from multiple places in the passage. Otherwise, our approach is the same: by restating the relationship between American and British railroads, we can determine the answer choice to which that same relationship applies.

## Applying the Method

*Passage Excerpt*

**[P1]** Locomotives were **invented in England**, with the first major railroad connecting Liverpool and Manchester in 1830. **However, it was in America that railroads would be put to the greatest use** in the nineteenth century.

**[P3]** While that speed was achieved under tightly controlled conditions, it still speaks to the **great disparity in railroad development between the United States and other countries** since the mid-twentieth century. **British trains travel at speeds much higher than those in America, where both the trains themselves and the infrastructure to support them have simply been allowed to fall behind.**

**Step 1: Identify the Relationship Between Elements**

An initial comparison between American and British railroads is made in Paragraph 1. Although trains were **invented in England**, they were put to **greater use in America**.

However, Paragraph 3 introduces a second point of comparison that notes the **great disparity** between modern British and American railroads. In today's world, **British trains travel at speeds much higher than those in America, where both the trains themselves and the infrastructure to support them have simply been allowed to fall behind.**

Based on these two comparisons, we can identify the following relationship:

**Originally, American railroads outperformed the British ones, but now, British railroads outperform the American ones.**

More generally, this structure is:

**First, A outperformed B, but later, B outperformed A.**

### Step 2: Match Answer Choices to That Relationship

Based on the comparisons in the passage, we concluded that the relationship between American and British railroads can be represented as: **First, A outperformed B, but later, B outperformed A**.

We can see this same relationship reflected in **Choice A**, *An apprentice artist surpasses a master's early works but does not live up to the master's later works*. Like the original British trains, the master's works came first but were then outclassed. However, like modern American trains, the apprentice's works did not live up to the master's later productions. Therefore, the relationship between the master's and apprentice's artworks parallels the relationship between British and American railroads.

The closest other answer would be Choice C, *A rookie athlete sets a team record but in later seasons his record is broken by a newer player*. However, notice that in this scenario, it is a newer player who breaks the rookie's record, not the person whose record the rookie originally surpassed. Therefore, the structure of this scenario would be: First, A outperformed B, but later, C outperformed both A and B. By contrast, all the elements in **Choice A** match the elements of the relationship between American and British railroads, making it the correct answer.

For an alternative method of explanation based more specifically on passage evidence, you can view this question in the UWorld Qbank (sold separately).

Lesson 8.1
# Bringing It All Together

This book has been designed to give you a focused and effective way to prepare yourself for the CARS section of the MCAT. Given the challenges that CARS can present, we began by discussing the purpose of the CARS section and how your mindset can affect your success. One of your biggest takeaways should be the belief that you can achieve your goals through practice and perseverance, and by concentrating on the process rather than just the outcome.

We have also sought to better acquaint you with the types of passages you will likely encounter on the exam, while discussing the various structural and rhetorical components they commonly contain. The accompanying CARS Passage Booklet includes multiple passages with detailed annotations designed to help you recognize a passage's structure and thus contribute to fuller comprehension as you read.

Finally, the heart of the book teaches you specific strategies for approaching the different types of CARS questions, breaking down the three overarching CARS skills into a more granular and detailed set of subskills. Understanding these finer distinctions between types of questions gives you a clearer sense of how the CARS section works, facilitating your ability to reason in the ways expected of you. You may also find the subskills helpful for guiding your study, by showing you the types of questions that you find easier or more challenging.

We hope you have found this book useful in all of these ways, and that it has increased your confidence in preparing for the CARS portion of the MCAT.

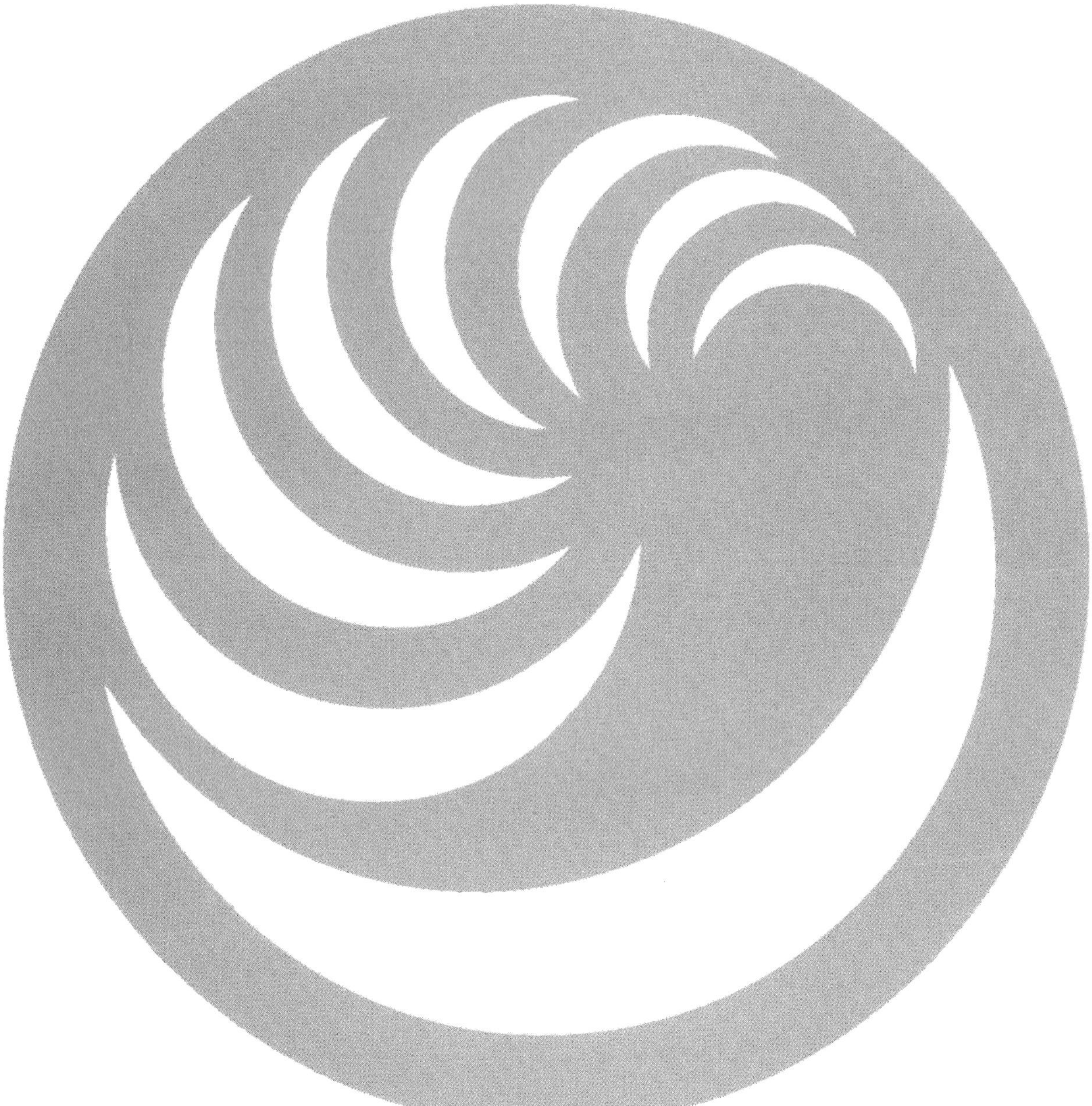